Applied Machine Learning for Smart Data Analysis

Computational Intelligence in Engineering Problem Solving

Series Editor: Nilanjan Dey

Techno India College of Technology, India

Proposals for the series should be sent directly to the series editor above, or submitted to:

Chapman & Hall/CRC
Taylor and Francis Group
52 Vanderbilt Avenue,
New York, NY 10017

Applied Machine Learning for Smart Data Analysis
Nilanjan Dey, Sanjeev Wagh, Parikshit N. Mahalle and Mohd. Shafi Pathan

Applied Machine Learning for Smart Data Analysis

Edited by

Nilanjan Dey
Sanjeev Wagh
Parikshit N. Mahalle
Mohd. Shafi Pathan

CRC Press
Taylor & Francis Group
Boca Raton London New York

CRC Press is an imprint of the
Taylor & Francis Group, an **informa** business

CRC Press
Taylor & Francis Group
6000 Broken Sound Parkway NW, Suite 300
Boca Raton, FL 33487-2742

© 2019 by Taylor & Francis Group, LLC

CRC Press is an imprint of Taylor & Francis Group, an Informa business

No claim to original U.S. Government works

Printed on acid-free paper

International Standard Book Number-13: 978-1-138-33979-8 (Hardback)

Library of Congress Cataloging-in-Publication Data

Names: Dey, Nilanjan, 1984- author. | Wagh, Sanjeev, author. | Mahalle, Parikshit N., author. | Pathan, Mohd. Shafi, author.
Title: Applied machine learning for smart data analysis / Nilanjan Dey, Sanjeev Wagh, Parikshit N. Mahalle and Mohd. Shafi Pathan.
Description: First edition. | New York, NY : CRC Press/Taylor & Francis Group, 2019. | Series: Computational Intelligence in Engineering Problem Solving | Includes bibliographical references and index.
Identifiers: LCCN 2018033217| ISBN 9781138339798 (hardback : acid-free paper) | ISBN 9780429440953 (ebook)
Subjects: LCSH: Data mining–Industrial applications. | Electronic data processing. | Decision support systems. | Machine learning–Industrial applications. | Internet of things. | Systems engineering–Data processing.
Classification: LCC QA76.9.D343 D49 2019 | DDC 006.3/12–dc23
LC record available at https://lccn.loc.gov/2018033217

Visit the Taylor & Francis Web site at
http://www.taylorandfrancis.com

and the CRC Press Web site at
http://www.crcpress.com

Contents

Section III Machine Learning in IoT............... 153

Section IV Machine Learning in Security 183

Preface

Applied Machine Learning for Smart Data Analysis discusses varied emerging and developing domains of computer technology. This book is divided into four sections covering machine learning, data mining, Internet of things, and information security.

Machine learning is a technique in which a system first understands information from the available context and then makes decisions. Transliteration from Indian languages to English is a recurring need, not only for converting out-of-vocabulary words during machine translation but also for providing multilingual support for printing electricity bills, telephone bills, municipal corporation tax bills, and so on, in English as well as Hindi language. The origin for source language named entities are taken as either Indo-Aryan-Hindi (IAH) or Indo-Aryan-Urdu (IAU). Two separate language models are built for IAH and IAU. GIZA++ is used for word alignment.

Anxiety is a common problem among our generation. Anxiety is usually treated by visiting a therapist thus altering the hormone adrenaline in the body with medication. There are also some studies based on talking to a chatbot. Cognitive Behavioral Therapy (CBT) mainly focuses on user's ability to accept behavior, clarify problems, and understanding the reasoning behind setting goals. Anxiety can be reduced by detecting the emotion and clarifying the problems. Natural conversation between humans and machines is aimed at providing general bot systems for members of particular organizations. It uses natural language processing along with pattern matching techniques to provide appropriate response to the end user for a requested query. Experimental analysis suggests that topic-specific dialogue coupled with conversational knowledge yields maximum dialogue sessions when compared with general conversational dialogue.

We also discuss the need for a plagiarism detector system i.e. Plagiasil. Some of our research highlights a technical scenario by predicting the knowledge base or source as local dataset, Internet resources, online or offline books, research published by various publications and industries. The architecture herewith highlights on scheming an algorithm, which is compliant to dynamic environment of datasets. Thus, information extraction, predicting key aspects from it and compression or transforming information for storing and faster comparison are addressed to explore research methodology.

Data summarization is an important data analysis technique. Summarization is broadly classified into two types based on the methodology: semantic and syntactic. Clustering algorithms like K-means algorithms are used for semantic summarization such as. Exploratory results using Iris dataset show that the proposed modified k-means calculation performs better than K-means and K-medoids calculation. Predicting the performance of a student is a great concern to higher education management. Our proposed system attempts to apply education data mining techniques to interpret students' performance. By applying C4.5 decision tree algorithm, the results get thoroughly analyzed and easily perusable.

Text summarization is one of the most popular and useful applications for information compression. Summarization systems provide the possibility of searching the important keywords of the texts and so that the consumer expends less time on reading the whole document. We study various existing techniques with needs of novel multi-document summarization schemes.

In the Internet of Things, smart cities need smart museums, as cultural interactivity is an important aspect of any nation. VANET (Vehicular Adhoc Network) provide Intelligent Transportation System (ITS). There has been an increasing trend in traffic accidents and congestion in the past years. So, advanced technological solutions have been proposed in an attempt to reduce such mishaps and improve traffic discipline. In the proposed system, we implement an android system that alerts the driver with messages. There is a need to prioritize message processing at node level. We address a novel scheme for assigning priority to messages and forwarding messages as per assigned priority. Time required to reach the destination is reduced due to proposed scheme.

Security is a very important concern in information computing systems. It can be addressed using various machine learning techniques. Spam emails present a huge problem. It is of utmost importance that the spam classifier used by any email client is as efficient as possible because it could potentially help not just with clearing out clutter and freeing up storage space but also blot out malicious emails from the eyes of layperson users. As email spam classification needs can vary for different types of usages, a comparative analysis helps us observe the performance of various algorithms on various parameters such as efficiency, performance and scalability.

Mobile malware is a malicious software entity, and it is used to disrupt the mobile operations and the respective functionalities. We propose machine-learning based classification of android applications to decide whether these applications are malware or normal applications. Malware detection technique by permission analysis approach and package malware setup is also presented.

Editors

Nilanjan Dey, PhD, earned his PhD from Jadavpur University, India, in 2015. He is an Assistant Professor in the Department of Information Technology, Techno India College of Technology, Kolkata, India. He holds an honorary position of Visiting Scientist at Global Biomedical Technologies Inc., California, USA, and Research Scientist of Laboratory of Applied Mathematical Modeling in Human Physiology, Territorial Organization of Scientific and Engineering Unions, Bulgaria. He is an Associate Researcher of Laboratoire RIADI, University of Manouba, Tunisia. He is also the Associated Member of University of Reading, UK, and Scientific Member of Politécnica of Porto, Portugal.

His research topics are medical imaging, data mining, machine learning, computer aided diagnosis etc. He is the Editor-in-Chief of the *International Journal of Ambient Computing and Intelligence* (IGI Global), US, *International Journal of Rough Sets and Data Analysis* (IGI Global), US, the *International Journal of Synthetic Emotions* (IGI Global), US, (Co-EinC) and *International Journal of Natural Computing Research* (IGI Global) (Co-EinC), US. He is Series Co-Editor of Advances in Ubiquitous Sensing Applications for Healthcare (AUSAH), Elsevier, Series Editor of Computational Intelligence in Engineering Problem Solving, CRC Press, and Advances in Geospatial Technologies (AGT) Book Series, (IGI Global), US, Executive Editor of *International Journal of Image Mining* (IJIM), Inderscience, Associated Editor of *IEEE Access* and *International Journal of Information Technology*, Springer. He has 20 books and more than 250 research articles in peer-reviewed journals and international conferences. He is the organizing committee member of several international conferences including ITITS, W4C, ICMIR, FICTA, ICICT.

Professor Sanjeev Wagh, PhD, is a Professor in the Department of Information Technology in the Government College of Engineering, Karad. He completed his BE in 1996, an ME in 2000 and a PhD in 2008 in Computer Science and Engineering at the Government College of Engineering, Pune & Nanded.

He was a postdoctoral fellow at the Center for TeleInFrastructure, Aalborg University, Denmark during 2013-14. He also completed an MBA from NIBM in 2015 in Chennai. His areas of research interest are natural Science computing, tnternet technologies and wireless sensor networks. He has written 71 research papers, published in international and national journals, and conferences. Currently five research scholars are pursuing PhD under his supervision in various universities. He is a fellow member of ISTE, IETE, ACM and CSI. He is editor for four international journals in engineering and technology.

 **Dr. Parikshit N. Mahalle** earned his BE degree in computer science and engineering from Sant Gadge Baba Amravati University, Amravati, India, and an ME in computer engineering from Savitribai Phule Pune University, Pune, India (SPPU). He completed his PhD in computer science and engineering (specializing in wireless communication) from Aalborg University, Aalborg, Denmark. He has more than 16 years of teaching and research experience.

Dr. Mahalle has been a member of the Board of Studies in Computer Engineering, SPPU. He is a member of the Board of Studies Coordination Committee in Computer Engineering, SPPU. He is also a member of the Technical Committee, SPPU. He is an IEEE member, ACM member, and a life member of CSI and ISTE. He is a reviewer for the *Journal of Wireless Personal Communications* (Springer), a reviewer for the *Journal of Applied Computing and Informatics* (Elsevier), a member of the Editorial Review Board for IGI Global – *International Journal of Ambient Computing and Intelligence* (IJACI), a member of the Editorial Review Board for *Journal of Global Research in Computer Science*, a reviewer for IGI Global – *International Journal of Rough Sets and Data Analysis* (IJRSDA), an Associate Editor for IGI Global – *International Journal of Synthetic Emotions* (IJSE), and Inderscience *International Journal of Grid and Utility Computing* (IJGUC). He is also a member of the technical program committee for international conferences and symposia such as IEEE ICC – 2014, IEEE ICACCI 2013, IEEE ICC 2015 – SAC-Communication for Smart Grid, IEEE ICC 2015 – SAC-Social Networking, IEEE ICC 2014 – Selected Areas in Communication Symposium, IEEE INDICON 2014, CSI ACC 2014, IEEE GCWSN 2014, GWS 2015, GLOBECOMM 2015, ICCUBEA 2015, ICCUBEA 2016. He has published 56 research publications at national and international journals and conferences with 177 citations. He has authored seven books: *Identity Management for Internet of Things*, River Publishers; *Identity Management Framework for Internet of Things*, Aalborg University Press; *Data Structures and Algorithms*, Cengage Publications; *Theory of Computations*, Gigatech Publications; *Fundamentals of Programming Languages – I*, Gigatech

Publications; *Fundamentals of Programming Languages – II*, Gigatech Publications; *Design and Analysis of Algorithms: A Problem Solving Approach*, (in press) – Cambridge University Press.

Dr. Mahalle is also the recipient of Best Faculty Award by STES and Cognizant Technologies Solutions. He has also delivered an invited talk on "Identity Management in IoT" to Symantec Research Lab, Mountain View, California. Currently he is working as Professor and Head of Department of Computer Engineering at STES's Smt. Kashibai Navale College of Engineering, Pune, India. He has guided more than 100 undergraduate students and 20-plus post-graduate students for projects. His recent research interests include algorithms, Internet of Things, identity management and security.

Dr. Mohd. Shafi Pathan is a Professor at Smt. Kashibai Navale College of Engineering, Pune. He completed his PhD (CSE) from JNTU Anantapur, India. He has completed a university funded research project on "Public key cryptography for cross-realm authentication in Kerberos." He has worked as the resource person for workshops and seminars. He is a reviewer for many national and international journals and conferences. He has worked as Head of the Publicity Committee for International Conference at the Global ICT Standardization Forum for India. He has worked as organizing secretary for an international conference on Internet of Things, Next Generation Networks and Cloud Computing, held at SKNCOE, Pune in 2016 and 2017. He was guest editor for a special issue of ICINC 2016 by IGI Global *International Journal of Rough Data Sets and Analytics*. He has authored four books titled as Fundamentals of computers, Computer Forensic and Cyber Application, Software Engineering & Project Management, Human – Computer Interaction and one book chapter titled Intelligent Computing, Networking, and Informatics by Springer, and he has published more than 40 research articles in national and international journals. He is a life member of ISTE and CSI.

Contributors

Aakash Atul Alurkar
Department of Computer
  Engineering
Smt. Kashibai Navale College of
  Engineering
Pune, India

Rohan Aswani
Department of Computer
  Engineering
Smt. Kashibai Navale College of
  Engineering
Pune, India

Jagdish W. Bakal
S S. Jondhale College of
  Engineering
Dombavali, India

Anjan Biswas
Department of Physics
Chemistry and Mathematics
Alabama A&M University
USA
and
Department of Mathematics and
  Statistics
Tshwane University of
  Technology
Pretoria, South Africa

Pranav Bhosale
Department of Computer
  Engineering
Smt. Kashibai Navale College of
  Engineering
Pune, India

Yogeshwar Chaudhari
Department of Computer
  Engineering

Smt. Kashibai Navale College of
  Engineering
Pune, India

P. S. Desai
Department of Computer
  Engineering
Smt. Kashibai Navale College of
  Engineering
Pune, India

M. L. Dhore
Vishwakarma Institute of
  Technology
Savitribai Phule Pune University
Pune, India

Sangram Gawali
Department of Computer
  Engineering
Bharati Vidyapeeth (Deemed to
  be University)
College of Engineering
Pune, India

Sachin P. Godse
Department of Computer
  Engineering
Smt. Kashibai Navale College of
  Engineering
Pune, India

Sanchit Gupta
Department of Computer
  Engineering
Smt. Kashibai Navale College of
  Engineering
Pune, India

Shaikh Naser Hussain
Department of Computer Science

College of Science & Arts
Al Rass AL Azeem University
Saudi Arabia

Rushikesh Jain
Department of Computer
  Engineering
Smt. Kashibai Navale College of
  Engineering
Pune, India

Sameer Joshi
Department of Computer
  Engineering
Smt. Kashibai Navale College of
  Engineering
Pune, India

Shashank D. Joshi
Department of Computer
  Engineering
Bharati Vidyapeeth (Deemed to
  be University)
College of Engineering
Pune, India

Shreeya Vijay Joshi
Department of Computer
  Engineering
Smt. Kashibai Navale College of
  Engineering
Pune, India

Salam Khan
Department of Physics
Chemistry and Mathematics
Alabama A&M University
Normal, Alabama, USA

Sachin M. Kolekar
Department of Computer
  Engineering
Zeal College of Engineering and
  Research
Pune, India

Shilpa G. Kolte
University of Mumbai
Mumbai, India

Kshitij Kulkarni
Department of Computer
  Engineering
Smt. Kashibai Navale College of
  Engineering
Pune, India

Qiuchi Li
Department of Information
  Engineering
University of Padua
Padua, Italy

Parikshit N. Mahalle
Department of Computer
  Engineering
Smt. Kashibai Navale College of
  Engineering
Pune, India

Simona Moldovanu
Faculty of Sciences and
  Environment
Modelling & Simulation
  Laboratory
Dunarea de Jos University of
  Galati
Galati, Romania
and
Department of Computer Science
  and Engineering
Electrical and Electronics
  Engineering
Faculty of Control Systems
Computers Dunarea de Jos
  University of Galati
Galati, Romania

Luminița Moraru
Faculty of Sciences and
  Environment

Modelling & Simulation
  Laboratory
Dunarea de Jos University of
  Galati
Galati, Romania

Pinakin Parkhe
Department of Computer
  Engineering
Smt. Kashibai Navale College of
  Engineering
Pune, India

Ketaki Pathak
Department of Computer
  Engineering
Smt. Kashibai Navale College of
  Engineering
Pune, India

Mohd. Shafi Pathan
Department of Computer
  Engineering
Smt. Kashibai Navale College of
  Engineering
Pune, India

Varun Patil
Department of Computer
  Engineering
Smt. Kashibai Navale College of
  Engineering
Pune, India

Jia Qian
Department of Information
  Engineering
University of Padua
Padua, Italy

P. N. Railkar
Department of Computer
  Engineering

Smt. Kashibai Navale College of
  Engineering
Pune, India

Siddhesh Sanjay Ranade
Department of Computer
  Engineering
Smt. Kashibai Navale College of
  Engineering
Pune, India

Sourabh Bharat Ranade
Department of Computer
  Engineering
Smt. Kashibai Navale College of
  Engineering
Pune, India

P. H. Rathod
Vishwakarma Institute of
  Technology
Savitribai Phule Pune University
Pune, India

Harsh Rohila
Department of Computer
  Engineering
Smt. Kashibai Navale College of
  Engineering
Pune, India

Gitanjali R. Shinde
Centre for Communication
Media and Information
  Technologies
Aalborg University
Copenhagen, Denmark

Vidyasagar Sachin Shinde
Department of Computer
  Engineering
AISSMS's College of Engineering
Pune, India

Piyush A. Sonewar
Department of Computer
  Engineering
Smt. Kashibai Navale College of
  Engineering
Pune, India

Devendra Singh Thakore
Department of Computer
  Engineering
Bharati Vidyapeeth (Deemed to
  be University)
College of Engineering
Pune, India

Prayag Tiwari
Department of Information
  Engineering
University of Padua
Padua, Italy

Naman Verma
Department of Computer
  Engineering
Smt. Kashibai Navale College of
  Engineering
Pune, India

Section I

Machine Learning

1

Hindi and Urdu to English Named Entity Statistical Machine Transliteration Using Source Language Word Origin Context

M. L. Dhore and P. H. Rathod

Vishwakarma Institute of Technology, Savitribai Phule Pune University

CONTENTS

1.1 Introduction

Out of vocabulary (OOV) words such as named entities (proper nouns, city names, village names, location names, organization names, henceforth referred as NE) are not a part of bilingual dictionaries. Hence, it is necessary to transliterate them in target language by retaining their phonetic features. Transliteration is a process of mapping alphabets or words of source language script to target language script by retaining their phonetic properties [1]. The choice language for our current proposal includes Hindi–English language pair in which source language is Hindi, which is the official national language of the Republic of India, and English, which is world's business language. Hindi uses the Devanagari script, which is highly phonetic in nature, while English uses the Latin script, which is less phonetic when compared with native Indian languages. In rest of the chapters, Hindi language and its associated Devanagari script will be referred to as Hindi, whereas the English language and Latin script will be referred to as English. According to official Act of India 1963, Hindi is the official language of India. After Chinese, English, and Spanish, Hindi is world's fourth most widely spoken language. Hindi is an Indo-Aryan language derived from Indo-European languages and spoken by half a billion people of India, Africa, Asia, America, Oceania, and Europe. According to a recent census, about 41.03 percent of the population in India speaks Hindi. Urdu is an Indic language spoken by approximately 52 million people in the southern, central, northern and western Indian states of Jammu and Kashmir, Telangana, Delhi, and Maharashtra. The Urdu literature has in its collections a wide range of poems and ghazals, which are also used in movie dialogs. A special case in the Urdu poetry is "shayari," which is a soulful way of expressing one's thoughts and ideas.

1.2 Hindi Phonology

For Hindi and other closely related languages, Panini's grammar had provided the phonetic classification of alphabets as vowels and consonants. They are commonly referred as vowel phonemes (Swara) and consonant phonemes (Vyanjanas) and will be denoted as V and C, respectively, in this chapter. These form a common base for all Indian languages. This classification helps to provide a unique encoding for the words in the languages based on Devanagari script. Sounds in Indian languages are divided into four groups, namely vowels, consonants, nasals and conjuncts. Vowels (V) are pure sounds. Consonants (C) are combination of one sound and a vowel. Nasals C(G)) are nasal sounds along with vowels and consonants. Conjuncts (CCV) are combination of two or more sounds. Indian languages are read as they are written because the script used by Indic languages is a syllabic alphabet representation. The syllabic alphabet is basically divided into Vyanjana (C) and Swara (V). Swara may be of two kinds, namely, short Swara or long Swara. Few graphical signs are used for denoting nasal consonants and are denoted by G.

1.3 Motivation

There is a need to work on transliteration generation problem if source language contains NE of multiple origins but share a common script. The

2001 census of India reveals that the Indian population largely consists of Hindus, following which the Muslims take the second place. In India, most of the documents require the name and address of the holder either in Hindi or English or both. Hindu and Muslim names when written and read in Hindi maintain their originality, but their transliterations differ in English. If the proper context in term of origin is not known, then there exist more chances of incorrect transliteration as mentioned in the original scenario. In India NE written in Hindi using Devanagari script may have different origins. For this work we have considered two origins: one is Indo-Aryan-Hindi (henceforth IAH) and the other is Indo-Aryan-Urdu (henceforth IAU). NE written in IAH follows the phonetic properties of Devanagari script for transliteration generation, whereas IAU follows the phonetic properties of Urdu. Urdu language follows perso-arabic script, and it is the national language of Pakistan. Urdu is also one of the official languages in India. NE of both IAH and IAU origin use the Devanagari Script which leads to the major concern in ambiguity. Consider as an example the name written in Devanagari "तौफीक" would be transliterated in English as "Taufik" by considering IAH context because it is written without using *nuqtas* for /फ़/ and /क़/. The correct transliteration is "Taufiq" according to IAU as /फ़/ and /क़/ are adopted to provide true phonemes of Urdu consonants in Hindi.

In the past, various utility bills including electricity, telephones and others in the state of Maharashtra were printed only in English, while other bills such as the Municipal Corporation Tax were printed in Hindi. In the recent years, these bills have names printed in Hindi as well as in English. However, the accuracy of names transliterated from English to Hindi and vice versa is not up to the mark in many cases. There are several reasons to this inappropriate transliteration problem. Few of them are multiple origins and one–many and many–one mapping due to unequal number of alphabets in the source and target language. Machine transliteration does not provide better accuracy as Devanagari script has two meanings for the English alphabet /a/ as /अ/ and /आ/. The stringent ambiguity is that the alphabet /अ/ is a short vowel and /आ/ is a long vowel in Devanagari. This ambiguity affects accuracy of transliteration generation especially from Roman to Devanagari. The correct name of the author of this chapter is /Manikrao//माणिकराव/, but it is printed as /मणिकराव/ on the electricity bill as the earlier records were in English and now have been transliterated in Hindi using machine learning. This chapter proposes Hindi to English transliteration by considering the origin of named entity in source language. In order to detect the origin, two separate data sets and two language models are used for IAH and IAU. Both the models are trained separately and the word alignment is obtained using GIZA++. Testing is done by calculating the probability of given named entity over both language models and higher probability between the two outcomes is taken as origin. If IAH has higher probability over the sequence of alphabets of given named entity then transliteration is carried out by IAH trained model, otherwise the IAU trained model is used.

1.4 Previous Work

Few of the phoneme-based approaches include the work of the authors in [2]–[6] and [8]; in [2] the author has used a combination of neural networks and knowledge-based systems to transliterate phone book entries in Arabic to English using a fixed set of rules. In [3] and [4], authors have used a weighted finite-state transducers (WFST) based on source channel method (SCM) to obtain back-transliteration for Japanese to English pair. In [5] the author has used statistical approach for transliteration using phoneme-based alignment for English–Chinese pair. Saha Sujan Kumar, Ghosh P S, Sarkar Sudeshna and Mitra Pabitra [6] have transliterated a gazetteer list in English to Hindi using a two-phase transliteration methodology, wherein an intermediate step was carried out to preserve the phonetic properties. Dhore modeled the "Hindi to English" and "Marathi to English" machine transliteration by using two weights on the basis of weight of the syllable. Much work has already been done by many researchers related to the grapheme-based approaches. We have considered the work related to the grapheme model and Statistical Machine Translation (SMT) for our study. In [9] the authors have developed an SMT model in which unsupervised learning and automatic training on a bilingual NE has been used. Nasreen and Leah (2003)[10] have proposed generative SMT using selected n-grams as a language model and GIZA++ for alignment of the Arabic–English pair. Grapheme-based model has gained popularity after the proposal of the Joint Source Channel Model (JSCM) in 2004 [11]. This model calculates the joint probability over source and target names. In [12] the author has modeled a language independent of SMT using Hidden Markov Model and Conditional Random Field. HMM together with Expectation Maximization is used to maximize the probability of transliteration for the Hindi and English pairs, while CRF is used to calculate the predicate maximum posterior probability. In [13] the author has improved the transliteration accuracy by detecting the origin of the word. Here, two different language models have been used to detect the origin, that is, CRF for training and a re-ranking model based on lexicon-lookup. In [14] the authors have used a 5-gram model and a Maximum Entropy Markov Model (MEMM) for alignment in order to improve the accuracy for the English–Persian pair. In [15], the authors have presented a Phrase Based -Statistical Machine Transliteration with a heuristic and a syllable segmentation mechanism, using the rule-based method for the English–Chinese pair. In [16], the authors have presented a back-propagation algorithm of neural networks for character recognition. In [17], the authors have proposed a statistical transliteration using two monolingual language models for Hindi and Urdu, and two alignments using GIZA++ for the Hindi–Urdu pair. In [18], the authors have proposed a mechanism to achieve the transliteration from Perso-Arabic to Indic Script using hybrid-wordlist-generator (HWG). In [19], the authors have presented the Beijing Jiaotong University -Natural Language Processing system of transliteration using multiple features. In [20], the authors have proposed a system using two baseline models for alignment between Thai Orthography and Phonology. This chapter proposes an SMT-based Devanagari-to-Roman transliteration by separately handling two disjoint data sets for two origins, two language models and separate training for each language. The same mechanism was proposed by Mitesh M. Khapra for Indic and western origin in 2009.

1.5 IAH and IAU

Table 1.1 depicts Devanagari script to Latin script phonetic mapping for IAH and IAU in full consonant form. Full consonant form means each consonant is always followed by the vowel "a." For example, क/ka/ indicated as /ka/ is a mapping for IAH and /क़/qa/ is a mapping for IAU accordingly. Devanagari script consists of 14 vowels, of which 12 vowels are pure vowels and 2 vowels are adopted from Sanskrit. It has 45 consonants in which 33 are pure consonants, 5 are conjuncts, 7 additional consonants from other languages and 2 traditional symbols. Each of these consonants further evolves into 14 variations through the integration of 14 vowels. The seven additional consonants in Hindi /क़, ख़, ग़, ज़, ड़, ढ़, फ़/ are adopted to

TABLE 1.1

Devanagari Script with its Phonetic Mapping for IAH and IAU

Vowel and Matra	Pure consonants	Conjuncts and Symbols	IAU Origin Consonants
अ-a-No Matra/, /आ-A-ा/, /ए-E-े/, /इ-i- ि/, /ऐ-aî-/, /ई-ee-ी/, /ओ-oo-ो/, /उ-u_ु/, /औ –au-ौ/, /ऊ-U-ू/,/अं-am-ं/, /ऋ-Ru_ृ/, /ॠ-RU_ृ/ /अ:-aH- :/	/क-ka/, /ख-kha/, /ग-ga/, /घ-gha/, /ङ -nga/, /च-cha/, /छ-chha/, /ज-ja/, /झ-jha/, /ञ-nya/, /ट-Ta/, /ठ-Tha/, /ड-Da/, /ढ-dha/, /ण -Na/, /त-ta/, /थ-tha/, /द-da/, /ध-Dha/, /न-na/, /प-pa/, /फ-pha/, /ब-ba/, /भ-bha/, /म-ma/, /य-ya/, /र-ra/, /ल-la/, /व-va/, /श-sha/, /ष-Sha/, /स-sa/, /ह-ha/	/क्ष-ksha/, /ज्ञ-dnya/, /श्र-shra/, /द्य-dya/, /त्र-tra/, /श्री-shree/, /ॐ-om/	/ड़ -rha/, /ख़-khha/, /ग़-gaa/, /ज़-za/, /फ़-fa/, /ड़-Dxa/, /क़-qa/

provide true phonemes in Persian, Arabic and Urdu. These seven consonant have a "dot" in its subscript referred as nuqta, which indicates the difference in phonetics from that of the regular Devanagari consonants.

Muslim names are mostly written by using IAU origin consonants in Hindi, however, the practice of not using nuqta generates wrong transliteration. For example, the named entity "तौफकि" is transliterated as "Tauphik" by considering IAH context, but the correct transliteration must be "Taufiq" according to IAU. If /तौफकि/ is correctly written as /तौफ़कि़/ using IAU origin, the consonants will then be transliterated to /Taufiq/. Nowadays nobody makes the use of nuqta while writing names of IAU origin in Hindi, which is a major concern in generating correct transliteration for Muslim names written in Hindi. This chapter mainly focuses on this issue and a solution is provided by creating two separate data sets, different language models and different trainings.

1.6 Methodology

The overall system architecture for Hindi to English named entity transliteration using correct origin in source language is shown in Figure 1.1. To avoid confusion, two separate data sets are created for IAH and IAU for Hindu and Muslim named entities. For IAH, syllabification of Hindu names is done by using phonetic mappings for all consonants and vowels except the seven additional IAU origin consonants /क़, ख़, ग़, ज़, ड़, ढ़, फ़/. Few examples are:

कुमार गंधर्व	[कु	मा	र] [गं	ध	र्व]	[ku	ma	r] [gan	dha	rva]
भमिसेन जोशी	[भी	म	से	न] [जो	शी]	[bhi	m	se	n] [jo	shi]
गरिजिा देवी	[गि	रि	जा] [दे	वी]	[gi	ri	ja] [de	vi]

For IAU, syllabification of Muslim names is done by using phonetic mappings for all consonants and vowels with the seven additional IAU origin consonants /क़, ख़, ग़, ज़, ड़, ढ़, फ़/. In order to improve the accuracy, the names are included twice in the data set, i.e., with and without nuqta. A few examples are:

तौफ़कि़	[तौ	फ़ी	क़]	[tau	fi	q]
तौफकि	[तौ	फी	क]	[tau	fi	q]
इक़बाल	[इ| क़| बा| ल]	[i	q	ba	l]
इकबाल	[इ| क| बा| ल]	[i	q	ba	l]
रज़्ज़ाक़	[र	ज़्ज़ा	क़]	[ra	zza	q]
रज्जाक	[र	ज्जा	क]	[ra	zza	q]

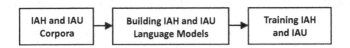

FIGURE 1.1
Overall System Architecture.

Syllabification is followed by constructing two separate language models for IAH and IAU. Language models are created using trigrams for both IAH and IAU. Language models are used to calculate the probabilities over the n-grams depending on their frequency in monolingual data set. For 1-grams, probabilities are calculated using unconditional probability while for 2-grams and 3-grams conditional probability is used.

In the following formulas s_i indicates syllable$_i$. Probabilities are calculated using the following formulas.

#Uni-gram Probabilities:

$$P(s_i) = [\text{count}(s_i)] / [\text{count}(\text{total number of 1-grams syllables in corpus})]$$

$$\text{Probability of syllable}_i = [\text{Frequency of syllable}_i \text{ in corpus}] / [\text{total number of syllables in corpus}]$$

#Bi-gram Probabilities:

$$P(s_i | s_{i-1}) = [\text{count}(s_{i-1}, s_i)] / [\text{count}(s_{i-1})]$$

$$\text{Probability of syllable}_{i-1} \text{ is followed by syllable}_i = [\text{Number of times syllable}_{i-1} \text{ followed by syllable}_i] / [\text{Frequency of syllable}_{i-1} \text{ in corpus}]$$

#Tri-gram Probabilities:

$$P(s_i | s_{i-2}, s_{i-1}) = \text{count}(s_{i-2}, s_{i-1}, s_i) / \text{count}(s_{i-2}, s_{i-1})$$

$$\text{Probability of syllable}_{i-1} \text{ is followed by syllable}_{i-2} \text{ and syllable}_{i-1} = [\text{Number of times syllable}_{i-2} \text{ followed by syllable}_{i-1} \text{ and syllable}_i] / [\text{Number of syllable}_{i-1} \text{ followed syllable}_{i-2} \text{ in corpus}]$$

The language model is followed by the transliteration model in which the mapping alignments of syllable sequences between Hindi and English NE are carried out. Transliteration models are trained by using three models of SMT. The test data set is a mixture of Hindu and Muslim names and is fully disjoint from trained data sets. During the decoding process, the named entity is given as an input to both language models in order to calculate probabilities over the syllables. This gives two probabilities over IAH and IAU language models and the higher probability between the two decides on which trained model is to be used for decoding.

1.7 Experimental Work

Figure 1.2 depicts the broader level experimental framework for our transliteration system. Statistical Machine Translation (SMT) is chosen for the experimentation as it also supports transliteration [19]. Moses is used as a platform for SMT, which in turn uses IRSTLM – la Ricerca Scientifica e Tecnologica Language Modelling and GIZA++. IRSTLM is used to build the language models for IAH and IAU. Language models are created for bi-grams, tri-grams, and four-grams. GIZA++ is used for alignment, which creates the Transliteration Model (TM). Three IBM models 1, 3, and 4 are used by GIZA++ for alignment, which in turn uses Expectation Minimization to give best possible alignments.

The overall methodology used for Hindi and Urdu to English Named Entity Statistical Machine Transliteration using Source Language Word Origin Context is depicted in Figure 1.3. For the SMT Moses toolkit is used. The overall logical flow of the system is implemented in three phases:

- Phase I: Normalization or pre-processing
- Phase II: Training
- Phase III: Testing

1.7.1 Normalization or Pre-processing Phase

In pre-processing phase the raw input data, in our case, the named entities need to be converted into system acceptable format. The set of NEs that cannot be processed directly by the system is known as raw input data. These NEs should be generalized in the system specified format. The Moses and GIZA++ toolkits accept and convert the input into syllabic units which is called as normalization or pre-processing. For the Moses

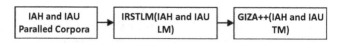

FIGURE 1.2
Broader Experiential Framework.

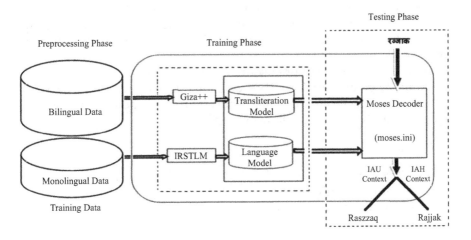

FIGURE 1.3
Overall System Architecture.

toolkit raw input data in Devanagari needs to be converted into syllabic format of the source language. In the proposed approach, one phonetic unit of the source language is considered as one syllabic unit. Two files are created for source and target languages and the parallel corpus is transformed into a numeric format. Few things which need to be taken care of while preparing the data set are as follows.

- High-quality phonetic syllable units should be on a single line
- Empty lines should not be there in the corpus, and lowercase alphabets should be used in order to speed up the training. This process is carried out by splitting the job into two submodules as:
  o Syllabification
  o Alignment

Syllabification and alignment of named entities is the process of partitioning of NEs of source and target languages into transliteration units (TU) by keeping their phonetic mappings intact. TUs of the source and target languages are referred as phonetic units or syllabic units. For example, the named entity in Hindi language written in Devanagari script 'भीमसेन' (person name) is syllabified as 'भी+म+से+न' and its equivalent phonetic syllabification in English is written as "Bhi+m+se+n". Moses accepts an input as syllabic units. These syllabic units are also called as syllables. As the syllabic unit is a phonetic equivalent, the phoneme is one of the features of implementation even though it is not explicitly stated. The target language label is a phonetic equivalent of Hindi in English. Named entities in Hindi written in Devanagari scripts form the source language transliteration units (STUs) and the corresponding phonetic units in English written in Latin script form the target language translation units (TTUs). Following is the algorithm for syllabification:

Input: NE is named entity in source language Hindi.

//initially pointer marks to the first Source Transliteration Unit (STU).

For each character in NE(named entity) do// For the each character of named entity NE, execute following steps:

1. Check whether a character is a consonant or a vowel.
2. If it is a consonant then add to present STU.
3. Else if it is a vowel then
   a. Add it to present STU.
   b. Move pointer to mark next STU.

Endforeach

Output: Syllabification of Source Named Entity

1.7.2 Training Phase

The training phase needs two things, one is the data set on which training has to be done and the other is the set of features on which the data set is to be trained. Parallel data obtained from the syllabification process is arranged in the required format of training in two separate files: one for source language and other for target language. These two files are maintained parallel to each other in terms of named entities syllabification units. Therefore, first file contains Hindi syllabified NEs and second file contains English syllabified named NEs. The IAH training data size was 20000 NE and was tested for 2200 NE including names of people, historical places, cities, town and villages. The IAU training data set size was 2200 NE and was tested for 400 NE including names of people, historical places, cities, town and villages. Following steps are used for undertaking training.

1.7.2.1 Language Model

LM software is invoked in this step. There are many language model tools compatible with Moses; few of them are IRSTLM, SRILM, IRSTLM, and CMU-Cam Toolkit. We have used IRSTLM Toolkit for language modelling. The IRSTLM uses algorithms and data structures to calculate, reserve and access very large n-gram language models. Generally, LM evaluation starts with grouping of n-grams and their frequency counters. It is followed by estimating smoothing parameters for each n-gram level and trimming of irregular n-grams. Finally, the LM file is created which contains probabilities of n-grams and back-offs weights.

IRSTLM provides a simple method to divide LM buildings into smaller and independent stages, so that it can be divided amongst independent processes. The process relies on a training script which uses very little of the computer's memory and applies the smoothing system. First, make a separate directory called working directory, where the script will save a lot of temporary files; after that, just run the script build-lm.sh. Figure 1.4 shows the screenshot of the Script for Building Language Model.

The LM is stored in the output file having an intermediate ARPA format as shown in Figure 1.5. The representation in ARPA format uses log probabilities to the base 10. In an ARPA file, the unigrams (1-grams) are unconditional probabilities while the n-grams with N of 2 or larger are conditional probabilities. In other words, when we see "–2.167781 the" in the 1-gram section, it is simply the log probability that the word {ma} will show up in the language, that is, in mathematical notation, P(ma), or "Probability of ma". But in the 2-grams section, when we see "–2.301030 of ma", this is the probability that {ma} will show up after of, i.e., in mathematical notation, P [ma/ni], or "Probability of {ma} given {ni}". There are three "special" words in a language

FIGURE 1.4
Script for Building Language Model.

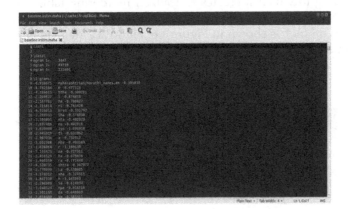

FIGURE 1.5
ARPA File.

model: <s>, </s>, and <unk>. The <s> denotes the beginning of a syllable, and the </s> denotes the end of a syllable. This differs from our original study where we had no concept of beginnings and ends of syllable—we just ignored syllable boundaries. The <unk> special word means "unknown". Consider examples {माणकिराव}, in the ARPA file probabilities generated for this माणकिराव on the basis of n-grams are as given here.

We know the total probability is always between 0 and 1 and not more than 1. Here, in the above example considering 1-gram log probabilities of {मा} is (−1.278754), <s> is (−1.102662) and </s> is (−0.465840). Now after taking their actual values of probabilities we should get the value 1. As 1-gram gives unconditional probabilities we need to add beginning and end of the syllable along with the probability value of {मा} in this case,

$$= 10^{(-1.102662)} + 10^{(-1.278754)} + 10^{(-0.465840)} = 0.07894 + 0.57893 + 0.3421 = 0.9978$$

From bi-gram onwards it uses conditional probability. Standard ARPA file format depicted in Figure 1.5 are the probabilities calculated for unigrams.

1.7.2.2 Transliteration Model

IRSTLM is compatible with GIZA++ and MGIZA++ Transliteration Model (TM). We used GIZA++ Toolkit for training language models for aligning the syllables using IBM models and HMM. GIZA++ requires good amount of RAM. The 4 GB and higher RAMs are always better as it is the most time-consuming step in the training process. TM calculates the probability of source syllable unit S for a given target syllable unit T, and it can be represented mathematically as P(T/S). Transliteration can be completed using syllable-by-syllable approach. To achieve this, Moses and word-aligner tool GIZA++ needs to be run for phrase extraction and scoring, corresponding to which it creates lexicalized reordering tables and finally generates Moses configuration file. The script for getting "phrase table" and "moses.ini file" is shown in Figure 1.6. Figure 1.7 depicts Phrase Table screenshot and Figure 1.8 shows screen shot for moses.ini.

FIGURE 1.6
Script for generating 'Phrase Table' and 'moses.ini file'.

FIGURE 1.7
Phrase Table.

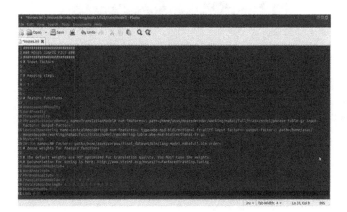

FIGURE 1.8
Moses.ini File.

GIZA++ includes:

- IBM-4 model for alignment
- IBM-5 model for dependency on syllables and smoothing
- HMM model for alignment
- Smoothing and distortion or alignment parameters
- IBM-1, IBM-2 and HMM models for improved perplexity calculation

The latest version of Moses software embeds calls to GIZA++ and mkcls software and hence there is no need to call them separately.

1.7.2.3 Testing Phase

Testing is done by calculating the probability of a given named entity over both language models and the higher probability between two is taken as the origin. If IAH has a higher probability over the sequence of alphabets of a given named entity then transliteration is carried out by IAH trained model otherwise IAU trained model is used.

Testing is the last phase in the system where the inputs NEs in Hindi having IAH and IAU origins are tested and the output is obtained in English by using the decoder. In our system, 2200 IAH and 400 IAU NEs are tested. If the output generated is correctly transliterated as expected, then the system has calculated accuracy based on only exact match. The main function of the Moses decoder is to find the highest scoring n-gram syllables TTUs in the target language corresponding to a given source named entity STUs as input. The decoder also provides the ranked list of the transliteration candidates.

FIGURE 1.9
Sample input.

FIGURE 1.10
Sample output.

FIGURE 1.11
Sample mismatch.

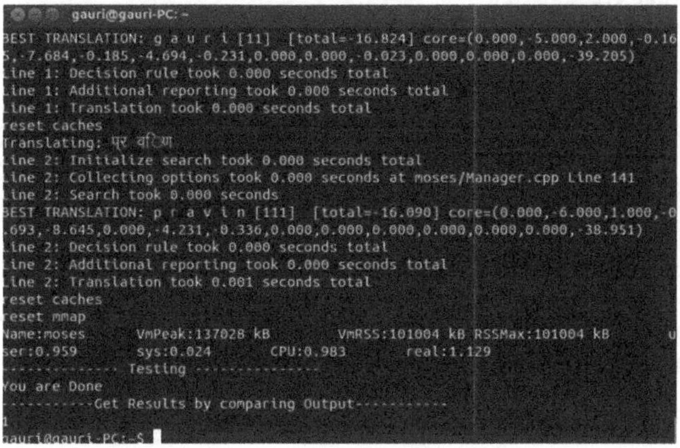

FIGURE 1.12
Console sample output.

The default setting weights used are: LM: 0.5, TM: 0.2, 0.2, 0.2, 0.2, Distortion Limit: 0.3.

Figures 1.9, 1.10, 1.11, and 1.12 shows sample input, sample output, sample mismatch and sample output testing at terminal.

1.8 Data

The training data sets for IAH and IAU are created from the authentic resources of the government websites. Few of them are census website, voter lists websites, road atlas books and telephone directories of Government of India in English and Hindi. IAH monolingual data set consists of 20000 NEs and IAU monolingual data set consists of 2200 unique named entities with only one candidate name in English. Test data for IAH were 2200 NEs and for IAU 400 NEs. Initial testing was done by using separate test data sets for IAH and IAU. Final testing was done by combining IAH and IAU test sets and it was 2600 NE.

1.9 Experimental Results

For the IAH training, data size was 20000 NE and tested for 2200 NE including names of people, historical places, cities, town and villages. For the IAU training, data set size was 2200 NE and tested for 400 NE including names of people, historical places, cities, town and villages. Accuracy is calculated for exact matches. Table 1.2 shows results of IAH

TABLE 1.2

Results IAH to English

For IAH	Bi-gram	Tri-gram	Four-gram
Training Set	20 K	20 K	20 K
Test Set	2.2 K	2.2 K	2.2 K
Exact Match Found	1521	1704	1865
Accuracy in %	69.13%	77.45%	84.77%

TABLE 1.3

Results IAU to English

For IAU	Bi-gram	Tri-gram	Four-gram
Training Set	2.2 K	2.2 K	2.2 K
Test Set	400	400	400
Exact Match Found	302	317	331
Accuracy in %	75.5%	79.25%	82.75%

TABLE 1.4

Results IAH, IAU to English

For IAH and IAU	Bi-gram	Tri-gram	Four-gram
Training Set	20 K and 2.2 K	20 K and 2.2 K	20 K and 2.2 K
Test Set	2600	2600	2600
Exact Match Found	1783	1867	1904
Accuracy in %	68.57%	71.80%	73.23%

to English. Table 1.3 shows results of IAU to English. Table 1.4 shows results of IAH, IAU to English accuracy, also known as Word Error Rate, which measures the correctness of the transliteration candidate produced by a transliteration system. Accuracy = 1 indicate exact match and Accuracy = 0 indicates no match for IAH and IAU. Following formula is used for calculating accuracy.

$$Accuracy = \frac{1}{N} \sum_{i=1}^{N} \begin{cases} 1, \textit{if correct match found} \\ 0, \textit{if incorrect match found} \end{cases}$$

where N is total number of NEs.

1.10　Conclusion and Future Scope

In this chapter, authors presented SMT-based approach to investigate transliteration problems for the multilingual support in printing electricity bills, telephone bills, municipal corporation tax bills, Aadhar card (unique identification number in India), and so on, in English as well as Hindi language when source language have the named entities with different context as either Indo-Aryan-Hindi (IAH) or Indo-Aryan-Urdu (IAU) for their phonetic mapping. The present-day system has many issues in the transliteration, and the information in both languages needs to be checked manually. Results achieved by our method are satisfactory and can be improved by increasing the training data set size for Muslim names. It is observed that approximately 10% named entities are wrongly classified in the combined input. This work can be extended by adding English names written in Devanagari in the source language as many addresses are a mix of Hindi, Urdu, and English named entities.

References

[1] Padariya Nilesh, Chinnakotla Manoj, Nagesh Ajay, Damani Om P. (2008), "Evaluation of Hindi to English, Marathi to English and English to Hindi", IIT Mumbai CLIR at FIRE.

[2] Arbabi M, Fischthal S M, Cheng V C and Bart E (1994) "Algorithms for Arabic name transliteration", IBM Journal of Research and Development, pp. 183-194.

[3] Knight Kevin and Graehl Jonathan (1997) "Machine transliteration", In Proceedings of the 35th annual meetings of the Association for Computational Linguistics, pp. 128-135.

[4] K. Knight and J. Graehl, "Machine Transliteration.' In *Computational Linguistics*, pp 24(4):599–612, Dec. 1998.

[5] Gao W, Wong K F and Lam W (2004), 'Improving transliteration with precise alignment of phoneme chunks and using contextual features", In Information Retrieval Technology, Asia Information Retrieval Symposium. Lecture Notes in Computer Science, vol. 3411, Springer, Berlin, pp. 106–117

[6] Saha Sujan Kumar, Ghosh P S, Sarkar Sudeshna and Mitra Pabitra (2008), "Named entity recognition in Hindi using maximum entropy and transliteration." Polibits (38) 2008, open access research journal on Computer Science and Computer Engineering. It is published by the Centro de Innovación y Desarrollo Tecnológico en Cómputo of the Instituto Politécnico Nacional, a public university belonging to the Ministry of Education of Mexico. Mexio, pp. 33-41

[7] Dhore M L, Dixit S K and Dhore R M (2012) "Hindi and Marathi to English NE Transliteration Tool using Phonology and Stress Analysis", 24th International Conference on Computational Linguistics Proceedings of COLING Demonstration Papers, at IIT Bombay, pp 111-118

[8] Dhore M L (2017) "Marathi - English Named Entities Forward Machine Transliteration using Linguistic and Metrical Approach", Third International Conference on Computing, Communication, Control and Automation 'ICCUBEA 2017' at PCCOE, Pune, 978-1-5386-4008-1/17©2017 IEEE

[9] Lee and J.-S. Chang. (2003),."Acquisition of English-Chinese Transliterated Word Pairs from Parallel-Aligned Texts Using a Statistical Machine Transliteration Model", In Proc. of HLT-NAACL Workshop Data Driven MT and Beyond, pp. 96-103.

[10] Nasreen Abdul Jaleel and Leah S. Larkey (2003), "Statistical transliteration for English-Arabic cross language information retrieval". In Proceedings of the 12th international conference on information and knowledge management. pp. 139–146.

[11] Li H, Zhang M and Su J (2004), "A joint source-channel model for machine transliteration", In Proceedings of ACL, pp.160-167.

[12] Ganesh S, Harsha S, Pingali P and Verma V (2008), "Statistical transliteration for cross language information retrieval using HMM alignment and CRF", In Proceedings of the Workshop on CLIA, Addressing the Needs of Multilingual Societies.

[13] M Khapra, P Bhattacharyya (2009), "Improving transliteration accuracy using word-origin detection and lexicon lookup" Proceedings of the 2009 NEWS

[14] Najmeh M N (2011), "An Unsupervised Alignment Model for Sequence Labeling: Application to Name Transliteration", Proceedings of the 2011 Named Entities Workshop, IJCNLP 2011, pp 73–81, Chiang Mai, Thailand, November 12, 2011.

[15] Chunyue Zhang et al (2012), "Syllable-based Machine Transliteration with Extra Phrase Features", ACL

[16] S P Kosbatwar and S K Pathan (2012), "Pattern Association for character recognition by Back-Propagation algorithm using Neural Network approach", International Journal of Computer Science & Engineering Survey (IJCSES), February 2012, Vol.3, No.1, 127-134

[17] M. G. Abbas Malik (2013), "Urdu Hindi Machine Transliteration using SMT", The 4th Workshop on South and Southeast Asian NLP (WSSANLP), International Joint Conference on Natural Language Processing, pp 43–57, Nagoya, Japan, 14-18 October 2013.

[18] Gurpreet Singh Lehal (2014), "Sangam: A Perso-Arabic to Indic Script Machine Transliteration Model". Proc. of the 11th Intl. Conference on Natural Language Processing, pp 232–239, Goa, India. December 2014. NLP Association of India (NLPAI)

[19] Dandan Wang (2015), "A Hybrid Transliteration Model for Chinese/English Named Entities", BJTU-NLP Report for the 5th Named Entities Workshop, Proceedings of the Fifth Named Entity Workshop, joint with 53rd ACL and the 7th IJCNLP, pp 67–71, Beijing, China, July 26-31, 2015, Association for Computational Linguistics

[20] Binh Minh Nguyen, Hoang Gia Ngo and Nancy F. Chen (2016), "Regulating Orthography-Phonology Relationship for English to Thai Transliteration", NEWS 2016

2

Anti-Depression Psychotherapist Chatbot for Exam and Study-Related Stress

Mohd. Shafi Pathan, Rushikesh Jain, Rohan Aswani, Kshitij Kulkarni, and Sanchit Gupta

Department of Computer Engineering, Smt. Kashibai Navale College of Engineering, Pune, Maharashtra, India

CONTENTS

2.1 Introduction

Anxiety and depression are two major public health issues prevailing in India. Around 5.6 million people in India suffer from either depression or anxiety, and 2.4 million of them are aged between 16 and 24 years. Most of these cases remain untreated because of the taboo associated with mental health care among the Indian population and also due to the higher cost incurred in treating these cases. This chapter aims at providing a solution to reducing anxiety and mild depression among the teens owing to study and exam pressure. Peer pressure and increased parental expectations in today's competitive world trigger anxiety [2] or mild depression among many teens, which can be treated with therapeutic sessions. With the increased pace in evolving human lives and changing environmental conditions, more and more people are being prone to depression. Anxiety is one of the precursors to depression. Anxiety is defined as a feeling of worry, nervousness, or unease about something with an uncertain outcome. With the increase in social media exposure and peer pressure, there are more number of cases of teenagers committing suicide because of insecurity and fear of separation. Anxiety and depression are topics that are still considered taboo in the Indian society. People do not talk about these issues openly. In addition, access to healthcare and the cost associated with it is also big issue. Mental health is not taken seriously. Teenagers growing up in emotionally weak households have low self-esteem [10] and are prone to develop symptoms of anxiety and depression from a very young age. Not being treated at the right time leads to severe depression. As we advance further in the 21st century, with changing environmental factors, we need to understand deeply the importance of mental health care and wellbeing. We suggest reducing anxiety using cognitive behavioral therapy (CBT). Every problem has an action and intent. The intent is the reason behind the action. The action would be the current user's problem. We apply CBT to change the intent of the user towards positive side from negative one. The user has to talk to the chatbot which determines the emotion of the user. The chatbot recognizes the user's problems and the reason behind it. The chatbot tries to solve the problems and continue to do so until it recognizes a positive response.

As soon as a text is received, the chatbot analyzes the emotion of the user [1, 3]. Another paper also mentions about psychiatric chat between human and computer [11].

2.2 Assumptions and Dependencies

The basic assumption made in this paper is that exam related anxiety is the only anxiety which is being addressed, it cannot address anxiety caused by other reasons. This is necessary to understand as anxiety can be caused by a variety of reasons. The physical and emotional symptoms of each of these anxieties differ and categorizing, addressing them is beyond the scope of this chapter.

In this chapter we propose using machine learning as a tool to address this real-world problem. Thus, the accuracy of this largely depends on the datasets which are being used to train the chat bot. The underlying requirement for the dataset to be a good dataset for this chatbot is where chat statements have good grammar English without abbreviations. Thus, when the chatbot replies it feels as if a human is replying. There is not an entire dependence on the neural network for response generation, the bot also uses natural language processing for searching the deepest phrase and getting the question context.

Thus, these assumptions are necessary to keep in mind as anxiety and depression are serious clinical and the bot needs to have abilities to comfort the human using the bot. These assumptions take care of the predefined environment which the bot the requires to operate.

2.2.1 Related Work

2.2.1.1 Probabilistic Topic Models

Probabilistic topic models concisely explain a large set of manuscripts with smaller number of groups/categories over the words. These categories are known as topics. First, semantic structure in the corpus is extracted by mining and similar words are removed using latent semantic analysis (LSA) [4], which uses a mathematical technique called singular value decomposition (SVD) to reduce the dimension of the data to make latent semantic space (LSS).

Probabilistic latent semantic examination (PLSA) [5] is an enhancement over the traditional LSA in which, instead of SVD, PLSA is based on a mixture of decompositions derived from latent class model (LCM). Latent Dirichlet allocation (LDA) [6] is another well-known method to categorize the words. BTM [3] is a broadened model of LDA, which distributes the content of the corpus by recognizing word co-word designs. A broad assessment has shown that it is a powerful model to learn with high characteristics over short reports.

2.2.1.2 Neural Networks

Neural networks are highly interconnected networks consisting of simple processing elements that dynamically process information based on external inputs and previous elements output. Feedforward neural network is one of the most popular neural network models in which the concept of multilayer perceptron is used. In this model, a back propagation technique is used to process information in the neural network. MLP forms a directed graph from the interconnections of the nodes. All nodes other than the input nodes have non-linear activation function. To decrease the disparity under various starting conditions, Fearing proposed a calculation by registering the weight's limits. Different techniques [5, 2] were proposed which revolve around information examination and earlier learning to ease the issue of angle vanishing. Bengio et al. proposed a strategy for greedy layer-wise training for profound systems to enhance the interpretability of concealed layers. Wan et al. associated neural systems with the various leveled subject model by utilizing neural systems to remove a trainable component. The entire neural network model coordinates the capacity of adapting low-level element change and abnormal state scene portrayal. Orchid et al. [11] incorporated subject space-based portrayal into a bolster forward neural system and connected it to discourse investigation. In any case, these cross-breed models were produced for ordinary information that contains adequate data (e.g. words in an individual document) primarily.

2.2.1.3 Social Emotion Detection

Social emotion detection anticipates the conglomeration of enthusiastic reaction shared by various clients, which started with the SemEval-2007 assignments. The SWAT framework, as one of the best performing frameworks on the "full of feeling content" assignment, was intended to misuse social feelings with singular words. Of late, Bao et al. proposed the emotion term (ET) algorithm and the emotion topic model (ETM) to connect social feelings with words and points separately. Rao et al. developed the multi-name regulated subject model (MSTM), the supposition inactive point display (SLIM) and the full of affective topic model (ATM) to distinguish social feelings in light of idle themes. The impediment of these models is that they require plenteous highlights to accumulate enough measurements.

2.3 Hybrid Neural Networks

The structure of neural systems, for example, the quantity of concealed neurons, is generally regarded as a black box. It is hard to give a reasonable semantic understanding to them. The probabilistic theme of concealed neurons can be better understood by the usage of subjects, i.e, for instance, the theme of "neuroscience" would give high probabilities to words like neurons, synaptic and hippocampal. Along these lines, we utilize the produced points as concealed neurons to improve the interpretability of neural systems. Before depicting the proposed model, we initially present the fundamental unit of our crossbreed neural systems, namely the idle semantic machine (LSM). Figure 2.1 exhibits a basic LSM that partners probabilistic point models with neural systems. LSM is made out of two distinct parts (or layers), which utilizes term-recurrence (TF) highlights and subjects as the info and yield, individually.

Picard proposed an idea of full feeling processing [3] as a part of human computer cooperation. Numerous specialists from software engineering,

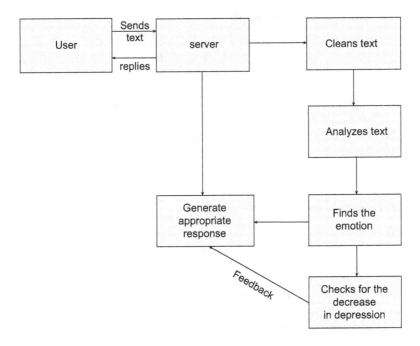

FIGURE 2.1
System architecture.

biotechnology, brain research, and subjective science are attracted to this area. From another perspective, investigation in the field of feeling location from printed information has further advanced to decide human feelings. Feeling location from content can be detailed mathematically as follows: Let A be the arrangement of all creators, T be the arrangement of every single conceivable portrayal of feeling communicating writings, and E be the arrangement of all feelings. Give r a chance to be a capacity to reflect feeling e of creator a from content t, i.e. r: A x T - >E, at that point the capacity r would be the response to the issue [4].

The idea of feeling acknowledgment frameworks lies in reality that, in spite of the fact that the meanings of T and E might be direct, the meanings of individual components, even subsets in the two arrangements of T and E, would be somewhat confounding. As the dialects are continually developing, new components may add on one side, for the set T. Due to the perplexing idea of human personalities, any feeling arrangements must be viewed as "names" clarified thereafter for various purposes. Although at present there are no standard groupings of "every single human feeling." Techniques utilized for content-based feeling acknowledgment framework [4, 5] are given next.

2.3.1 Keyword Spotting Technique

The watchword spotting strategy can be depicted as the issue of discovering events of catchphrases (cherish, outrage, happiness, bitterness, shock and dread) from a given content record. Numerous calculations to break down feeling have been recommended before. With regard to feeling recognition, this strategy depends on certain predefined watchwords. These feeling words are ordered into catchphrases, for example, disturbed, dismal, cheerful, irate, frightful, shocked, and so on. Events of these catchphrases can be found and in light of that a feeling class can be appointed to the content report.

2.3.1.1 Lexical Affinity Method

Recognizing feelings in view of related watchwords is simple to utilize and a direct technique. Watchword spotting method is reached out into lexical partiality approach, which allocates a probabilistic "fondness" for a specific feeling to subjective words separated from getting enthusiastic catchphrases. These probabilities are a piece of etymological corpora, which still have a few disservices likewise; the appointed probabilities are one-sided toward corpus-particular type of writings. Also, it passes up a great opportunity passionate substance that lives further than the word-level on which this procedure works, e.g. watchword "mischance" having been appointed a high likelihood of showing a negative feeling, would not contribute accurately to the passionate appraisal of expressions like "I met my sweetheart unintentionally" or "I avoided an accident."

2.3.1.2 Learning-Based Methods

Learning-based techniques are being utilized to dissect the issue in an alternate way. At first the issue was to decide the notion from input content information. However, at present, the issue is to order the information writings into various feeling classes. Unlike the catchphrase-based discovery strategies, learning-based techniques endeavor to perceive feelings in view of already prepared classifier, which apply hypotheses of machine adapting, for example, bolster vector machines [8] and restrictive irregular fields [9], to confirm that the information content has a place in the appropriate feeling class.

The following section represents some lexical resources that has been assembled by many researchers over the past few decades in order to support and provide effective computing along with the recently suggested methods.

2.3.1.3 Lexical Resources

In the list of benefits, an essential one described a rundown of 1,336 descriptive words that were named manually [1]. WordNet-Affect was presented as a progressive system of emotional area marks [2]. The subjectivity dictionary [3] involves more than 8,000 words. In [4], the idea was propelled by the presumption that distinctive faculties of a similar term may have diverse sentiment related properties. A vocabulary named SentiFul database is presented in [5]. Feelings detection approaches and emotion acknowledgment methodologies can be comprehensively grouped into keyword based, semantic principles based and machine learning systems. We have additionally recognized them in light of whether they utilize any influence dictionaries.

2.3.1.3.1 Approaches Based On Keyword Using Affect Lexicons

Keyword-based methodologies are connected at the essential word level [6]. Such a basic model cannot adapt to situations where impact is communicated by interrelated words.

2.3.1.3.2 Linguistic Rules-based Approaches

Computational linguists define a language structure by using many rules.

- Rule-based methodologies with influence dictionaries: The ESNA framework [7] was created to arrange feelings in news features.

Chaumartin [8] manually added seed words to feeling records and made a couple of standards in their framework UPAR 7, which distinguishes what is being said in regard to the fundamental subject, and lifts its feeling rating by using reliance diagrams. The impact of conjuncts [9] was contemplated utilizing rules over linguistic structure trees and lexical assets, for example, General Inquirer and WordNet. A standout among the latest run-based methodologies [10] can perceive nine feelings. The majority of these methodologies complete a fantastic assignment of characterizing decisions that unravel complex dialect structures. Be that as it may, planning and adjusting rules is a long way from a paltry undertaking. Also, approaches utilizing influence dictionaries experience the ill effects of the rigidity of taking into account feelings other than those effectively recorded.

- Rule-based methodologies without influence dictionaries: As another option to utilizing influence vocabularies, [11] in order to understand the fundamental semantics of dialect, an approach is proposed by the authors in [11]. Another fascinating methodology is to perceive feelings from content rich in figurative information. Albeit such techniques have the adaptability of utilizing any arrangement of feelings and are hence handier, the principles are particular to the portrayal of the source from which knowledge is taken.

2.3.1.3.3 Approaches to Machine Learning

To remove the restrictions looked by manage-based strategies; specialists conceived some measurable machine learning procedures, which can be subdivided into regulated and unsupervised systems.

- Supervised machine learning with influence dictionaries: One of the soonest regulated machine learning strategies was utilized by Alm, where they utilized a progressive consecutive model alongside SentiWordNet list for extensively filtered feeling grouping. Sentences from blogs are organized utilizing Support Vector Machines (SVM). Albeit, directed learning performs well; it has the unmistakable burden that vast clarified informational indexes are required for preparing the classifiers and classifiers prepared on one area for the most part don't perform so well on another.

- Supervised machine learning without influence vocabularies: A correlation among three machine learning calculations on a film survey informational collection presumed that SVM plays out the best. A similar issue was additionally endeavored utilizing the delta tf-idf work.

- Unsupervised machine learning with influence vocabularies: An assessment of two unsupervised strategies utilizing WordNet-Affect used a vector space display and various dimensionality lessening techniques. News features have been ordered utilizing straightforward heuristics and more refined calculations (e.g., comparability in an idle semantic space).

- Unsupervised machine learning without influence dictionaries: Some motivating work done here incorporates "LSA single word" which measures similitude amongst content and every feeling and the "LSA feeling synset" approach which utilizes WordNetsynsets. Our approach shares a comparative instinct as that of the "LSA feeling synset" strategy, but with some remarkable contrasts as we utilize Pointwise Mutual Information (PMI) to register the semantic relatedness, which is additionally improved by setting reliance rules. In spite of the fact that utilization PMI to assemble insights from three web search tools, they contrast a whole expression with only one feeling word because of long web-based preparing times, while, in our approach, each significant word with an arrangement of agent words for every emotion is taken into account along with its context.

Feeling is an intricate wonder for which no accurate definition considering all emotions has been recognized and given. However, the commonly used definition considers feeling as "a scene of interrelated and synchronized changes in the conditions of all or an extensive bit of the five organismic subsystems (Data planning, Support, Official, Activity, Screen) in light of the evaluation of an external or internal lift event as pertinent to noteworthy stresses of the living thing." The term feeling centers to a single section meaning the process of subjective experience and is in this way only a little bit of an inclination. Slants are not much specific, less remarkable loaded with feeling ponders, after effect of two estimations—imperativeness and weight. Supposition is an individual conviction or judgment that is not set up on affirmation or sureness [1]. These impact related wonders have generally been thought about through and through by orders, for instance, thinking or cerebrum science. In any case, on account of the advances in figuring and the enlarging piece of development in normal everyday presence, the earlier decades have shown an extending eagerness for building programming systems that will have a greater impact. On the whole, such structures, in order to benefit by the data secured in human sciences, interdisciplinary methodologies suggest the usage of the present theoretical models as explanation behind planning computational ones. This zone explains the bleeding edge in the three spaces. Our current research is solidly related to: approaches to manage feeling acknowledgment in synthetic intellectual competence, assessment models in cerebrum science, and learning bases in NLP application.

2.4 System Architecture

To perform an online psychotherapist chatbot system will start on the client side and go towards the server side and will end on the client side. Figure 2.1 shows the detailed system architecture. First, the user logs onto the system in which the user credentials are checked from the database on the server, after which every message is sent to the server where a response will be created and will be returned to the user.

2.4.1 Text Cleaner

The text cleaner performs the task of cleaning the input text. It waits for complete sentences from the user and then corrects grammatical errors. The text cleaner further breaks statements into words, [7] stems them with root words and removes stop words to form an array of stock words. These stock words act as input to the text analyzer which finds the emotions from text [4].

2.4.2 Text Analyzer

The text analyzer forms the basis for two things in the basic architecture. Figure 2.2 shows the structure of the context generation process. It makes up the context [8] of the human situations using a Tree. The text analyzer tries to find answers to the four questions in every context: "What are you

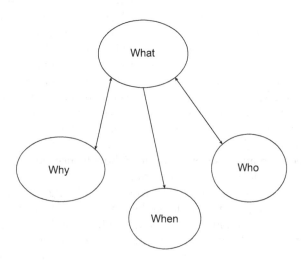

FIGURE 2.2
Context storage using a tree.

feeling?" "Why are you feeling?" "Who is making you feel this way?" "Since when do you feel this way?" These four "what, why, who, when" questions set up the basis of the situations. The text analyzer also analyzes emotions in the given array from the text cleaner.

2.4.3 Response Generation

CBT uses three attributes to change the human model of thinking: the situation, meaning and the conclusion. Every situation in the human brain is mapped to a conclusion it forms from its outcomes. These conclusions are highly dependent on the meaning one makes out from that situation. For any given situation, the CBT theory identifies every negative meaning and replaces it with a positive meaning which rewrites negative conclusions with a positive one.

We do this with the help of a neural network. Figure 2.3 shows the process of Text tellering mechanism. The neural network is trained to take in two inputs, which are the emotion class and the context of the situation. The neural network is trained to get the best possible positive meaning from the input negative meaning. All the situations and meanings are mapped and labeled with a class. The neural network outputs the best positive meaning for a given negative situation. The response obtained on identifying the type of output class generates the text for

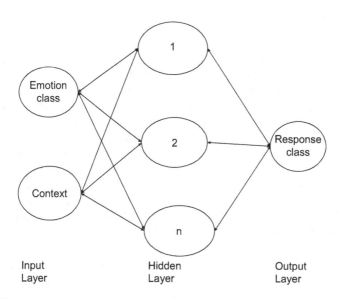

FIGURE 2.3
Neural network representation for output classification.

the given class. This response is sent back to the server and is given back to the user.

The emotion classifier continuously analyses input and hence gives the emotion feedback to the text generator to improve the type of response given back to the user. This increases the accuracy of matching the user context and thinking.

2.5 Functional Requirements

2.5.1 System Feature1

The chat client is an android application which acts as an intermediate between the server bot and the user. The chat client stores the previous history of chat with itself for further analysis of user emotions. It has an MCQ test to estimate the levels and causes of depression. The communication between the client and the server happens through JSON requests and responses. The chat client also does the text pre-processing of the chat message and makes it suitable for the chatbot to analyze and perform analysis on it. The client is also responsible for securing the connection and the data sent to it. The data sent to the chatbot has to be encrypted as it is very important for the data to remain in safe places.

2.5.2 System Feature2

Machine learning and information mining [9] regularly utilize similar strategies and cover fundamentally everything. However, machine learning centers gather information by using the properties that are known which are learned from the preparation information, information mining centers around the revelation of (beforehand) obscure properties in the information (this is the investigation venture of information disclosure in databases). Information mining utilizes numerous machine learning strategies, yet with various objectives. Machine adapting these strategies either utilizes information mining techniques such as "unsupervised learning" or is used as a preparing venture to enhance student precision. A significant part of the perplexity between these two research groups (which do regularly have isolate gatherings and separate diaries, ECML PKDD being a noteworthy exemption) originates from the fundamental suspicions they work with. In machine learning, execution is typically assessed based on the capacity to replicate known information, while in information revelation and information [5] mining (KDD) the key undertaking is the disclosure of previous obscure learning. While assessing for known information, a clueless (unsupervised) technique will effortlessly be beaten by other managed strategies, while in a normal KDD undertaking, direct techniques cannot be utilized because of the inaccessibility of preparing information. The bot server employs the knowledge produced by machine learning, KDD and natural language processing in designing the next best response to the user. Natural language processing is an important part of extracting knowledge from the user chat messages and then reproducing output in human understandable format. The bot server works on the intention–action pairs and produce alternate combinations of the intention–action pair so that CBT and MCBT can be applied, and the user thinking can be changed with these responses.

The chatbot tries to decrease anxiety and mild depression among teens at a very cheap cost. The chapter focuses on bringing psychiatry and technology together to take mental care to the next level, thereby improving the lives of people.

2.6 External Interface Requirements

2.6.1 Hardware Interfaces

Since the application is mostly software, the hardware interface requirement is very less. Nonetheless, the required hardware interfaces are:

Android device: The application is built as an Android application, hence an Android interface or a device is required to run the application. These devices can be an android mobile, android tablet or a device running an android emulator.

Hardware internet interfaces: The application runs only with the help of an internet connection. So, the internet interfaces that help the device or application connected to the application server are required. These may include Wi-Fi routers, modems switches, and so on.

2.6.2 Communication Interface

The user goes to the login page and authenticates themselves. The user can see his last interaction with the application.

After the successful authentication of the user, they can start interacting with the application. The user can generate the input for the application by typing the text and then the input is transferred to the application server.

The server takes the current input, processes it with the previous input of the user and the database built through interactions with other users and generates the appropriate output.

After the output is generated, it will be sent to the user. The user reads the generated output and reacts accordingly by typing another text. This text is then again given as input to the application server and the communication cycle carries on as long as the user continues to give input to the application server.

2.7 Non-Functional Requirement

2.7.1 Availability

Chatting applications are ubiquitous nowadays. People are online most of the time on chatting applications like WhatsApp, Hike, and others. A chatbot is an intelligent application with which a user can chat and get a reply as if from a human. The user may get time during any time of the day to chat so it is made available 24*7.

2.7.2 Backup

In a real-time system no issues in the system should affect the user. To give a continuous service, the system should be regularly backed up. So, if anything happens to the database, the system should switch to the backup database and continue to provide service. Also, when the original database is available the backup should store the current status in it (updated frequently).

2.7.3 Fault Tolerance

The chatbot will be used by depressed people, hence if the system fails due to some technical reasons it might lead to making the user more stressed. Therefore, care should be taken to make the system fault tolerant.

2.7.4 Performance

The response time is a critical measure for any application because people use the application to make their world simple, better, and to save time. Hence the response time should be as short as possible. The response time depends on many factors such as number of users that can be connected to the server at any given instance. This can be done by running a background process on many servers in order to address the number of users at a time, since the number of people using the system keeps changing. The connectivity issue can be solved by minimizing the data transfer, i.e. by preprocessing the string at the user end and passing only the required characters to the server side.

2.7.5 Resource Constraints

Nowadays mobile phones have good storage space and processing units, but the previous versions of mobile have very minimal resources. As the application is being used by people in stress or depression, it is highly possible that this category of people might also be financially weak who could not afford high-end mobile phones. So, the system must be equipped to run on low-end phones as well.

2.7.6 Security Requirements

The people using the application would mostly be under stress or anxiety. So, the chatting text might contain any sensitive or personal information which may expose the user's personal life. The information must be protected from the outside world. In order to do this, an authentication would be made each time user opens the application where the username and password will be asked to check that it is the same person using the application.

2.7.7 Safety Requirements

If there is extensive damage to the database due to failures such as power failure or the storage crashes, the backup database should kick in and the application server should respond to the users using this backup database. The backup database should be updated to the current state by redoing the committed transaction from the backed-up log, up to the time of failure so that the chat will be more relevant to the user. Also, in case of unavailability of the database the backup should be used by the server.

In the meantime, the failed database should attempt its recovery. After the database is successfully recovered it should be kept as a backup database in case of any failure or damage to the database currently in use.

2.7.8 Security Requirements

The user account should be protected from outside attacks or hacking as the information is very sensitive and might cause harm to the user. For this a login system has been implemented in which user has to enter his username and password to login and access the application. This can be termed as client-level security.

The data can also be intercepted when the user is chatting with the server. If the data is transferred in plaintext format the data can be easily understood by the interceptor giving a clear picture of the user's state of mind and could be used to cause harm. The interception can be provided by starting a VPN between the user and server. Security can further be enhanced by encrypting the sent and received data from and to the user using various encryption algorithms like Triple DES and RSA. The information can be decrypted by the application and the server using the decryption key. This is transport-level security.

There is also a security concern for the server and the database as the attacker can modify the responses sent to the user which could psychologically break down the user. Also, the sensitive information of the user could be stolen. This can be avoided by employing good and trusted storage partners.

2.8 System Requirement

2.8.1 Database required

Whenever a person downloads the app, he/she needs to sign up in the system; thereafter every time they open the app they need to log in. This functionality maintains the privacy of the conversations with therapist; the information in the chatbot is private. Login information of the user is saved in the database and verified every time the user logs in. This database is useful for connecting the user to the chatbot.

Before starting the chat between the user and the chatbot, multiple choice question test is taken to determine the emotion of the user, which is sent to the server and saved in the user database information. This database is accessed while chatting with the user. As the information of the user and emotions are single fixed quantities, relational databases are used. There are many relational databases for use and, in this application, MySQL is used to store the information. MySQL is required to be installed on the Windows Server 2016 with proper path configuration.

Other important things are datasets. Datasets are used to derive an equation which is later used for determining the emotion [7] while creating a response to the user. One of the datasets to determine the emotion is ISEAR dataset. Various datasets are available to create an equation. ISEAR is used here to create an equation to determine the emotion of the user. Datasets are also needed to be stored on the windows server.

2.8.2 Software requirements

Android platform is a mandatory requirement in order to use the chatbot application. The version of the Android operating system required by the user is 4.4.4, i.e. KitKat or above. Client requirements are simple, for example, android mobile and an internet connection.

Moving to the server side of the application, Windows Server 2016 is used with basic EC2 configuration with a minimum of 30 GB of space. The server is coded in Django which is a python library; hence python and Django need to be installed on the windows server. The server should be powerful to handle multiple requests sent by various clients using the application. Various types of datasets are required to develop an equation that would most appropriately decrease anxiety or mild depression of the user. ISEAR is one of the examples of the datasets that are used to determine the emotion of the user.

2.8.3 Hardware requirements

To give a response to multiple users with issues of anxiety and mild depression caused due to stress, a strong hardware with multithreading is needed. Each user who is chatting with the chatbot about their problems is assigned a separate thread.

User who needs to chat with the bot needs an android phone with following minimum requirements:

1. 512mb RAM

2. 512mb Disk space

3. Android (KitKat 4.4.4 or more)

4. An internet connection

2.8.3.1 System Implementation Plan

To build a psychotherapist chatbot to treat anxiety and mild depression, multiple layers of the application is required. These depressions can be caused because of studies, parental pressure, and so on. There are four basic layers of the whole application:

1. Front layer

2. Business

3. Logic

4. Database

Front layer is designed in android, which has the login page and the actual chatting application. The login page validates the user credentials from the database and if found correct the actual chatting page is displayed where the user can type the messages and send it to the server. Before starting the actual chat, a quiz is taken which determines the polarities of all emotions; the emotion with maximum polarity is considered as the current emotion of the user.

Business logic is the actual code which determines the actual emotion and generates a response to be given back to the user. When the quiz is taken, the answers are sent down to the server and the polarities of the all the emotions are identified. Once the emotion is determined the user starts talking to the chatbot. All the messages the user types are sent down to the server. The action and intent are determined for every problem that the user states. Response is generated such that the intent of every action turns from negative to positive.

Database is used to store the user's credentials, emotion, actions and intents. User's credentials are stored during the signup process and checked during login process. Emotions of the user are stored in the database. Actions and the intents are stored during the chat process.

First, a server is made in Django python that accepts the requests and messages; next, the android app is designed which contains login, signup, and the chat page. A database is created to save the credentials of the users, after which a quiz is taken to determine the emotion. The algorithm to determine the emotion based on the answers is made on the server side. The emotion is saved in the database against the username. The response generator is then created. From the problems keyed in by the user, action and intents are extracted and saved in the database. Response is generated using the algorithm on the server side, which tries to change the intent of every action. Once all the intents are changed, the emotion polarity is altered and the work is done.

2.8.4 Base Knowledge for Natural Language Processing Applications

Numerous NLP applications have been generated with the help of manually created knowledge repositories, such as WordNet, Cyc, ConceptNet, and SUMO. A few authors attempted learning ontologies and relations by utilizing sources that developed in time. For example, Yago utilizes Wikipedia to pluck out concepts utilizing standards and heuristics dependent on the Wikipedia classifications. Different ways to knowledge base population include semantic class learning and learning based on relations. DIPRE and Snowball mark a little collection of instances and make skillful examples to separate cosmology ideas.

2.9 Conclusion

The chapter discussed the uses of machine learning and natural language processing to successfully apply CBT and generate human-like responses by a chatbot.

References

[1] R. Cowie, E. Douglas-Cowie, N. Tsapatsoulis, G. Votsis, S. Kollias, "Recognition of Emotional States in Natural human-computer interaction," in IEEE Signal Processing Magazine, vol. 18(1), Jan. 2009.

[2] Parrott, W.G, "Emotions in Social Psychology," in Psychology Press, Philadelphia 2001

[3] C. Maaoui, A. Pruski, and F. Abdat, "Emotion recognition for human-machine communication", Proc. IEEE/RSJ International Conference on Intelligent Robots and Systems (IROS 08), IEEE Computer Society, Sep. 2008, pp. 1210-1215, doi: 10.1109/IROS.2008.4650870

[4] C.-H. Wu, Z.-J.Chuang, and Y.-C.Lin, "Emotion Recognition from Text Using Semantic Labels and Separable Mixture Models," ACM Transactions on Asian Language Information Processing (TALIP), vol. 5, issue 2, Jun. 2006, pp.165-183, doi:10.1145/1165255.1165259.

[5] Feng Hu and Yu-Feng Zhang, "Text Mining Based on Domain Ontology", in 2010 International Conference on E-Business and E-Government.

[6] Z. Teng, F. Ren, and S. Kuroiwa, "Recognition of Emotion with SVMs," in Lecture Notes of Artificial Intelligence 4114, D.-S. Huang, K. Li, and G. W. Irwin, Eds. Springer, Berlin Heidelberg, 2006, pp. 701-710, doi: 10.1007/11816171_87.

[7] C. Strapparava and A.Valitutti, "Wordnet-affect: an effective extension of wordnet," in Proceedings of the 4th International Conference on Language Resources and Evaluation, 2004.

[8] T. Wilson, J. Wiebe, and P. Hoffmann, "Recognizing contextual polarity in phrase-level sentiment analysis," in Proceedings of the Conference on Human Language Technology and Empirical Methods in Natural Language Processing, 2005, pp.347–354.

[9] A. Esuli and F. Sebastiani, "Sentiwordnet: A publicly available lexical resource for opinion mining," in Proceedings of the 5th Conference on Language Resources and Evaluation, 2006, pp. 417–422.

[10] A. Neviarouskaya, H. Prendinger, and M. Ishizuka, "Sentinel: Generating a reliable lexicon for sentimentanalysis," in Affective Computing and Intelligent Interactionand Workshops, 2009.

[11] Kyo-Joong Oh, Dongguan Lee, Byung-soo Ko, Ho-Jin Choi: "A Chatbot for Psychiatric Counseling in Mental Healthcare Service Based on Emotional Dialogue Analysis and Sentence Generation"

3

Plagiasil

A Plagiarism Detector Based on MAS Scalable Framework for Research Effort Evaluation by Unsupervised Machine Learning – Hybrid Plagiarism Model

Sangram Gawali, Devendra Singh Thakore, and Shashank D. Joshi
Department of Computer Engineering, Bharati Vidyapeeth (Deemed to be University), College of Engineering, Pune, India

Vidyasagar Sachin Shinde
Department of Computer Engineering, AISSMS's College of Engineering, Pune, India

CONTENTS

3.1 Introduction

In today's progressive technological era, new concepts, new methods, new algorithms and terms are being written every day, resulting in the growth of the information tree. It is critical to evaluate efforts of authors, which can be achieved by means of plagiarism detection. Plagiarism is nothing but copying of content and ideas of someone without attributing them [1]. In order to publish quality literature, it is important for publishers to check for plagiarism and evaluate the efforts of the author. Academic institutions face greater issues of plagiarism owing to various factors like social engineering and sharing of submissions. Plagiarism detectors thus play a major role in the task of effort evaluation [1].

Information sources are identified by appropriate reference specifications, and if such references are absent for a particular statement then it implies that it is the author's original effort. Some sources of information are the internet, books, journals, and so on. An honest document is one that provides attribution to original authors [2].

Some of the evident factors of plagiarism are as follows:

1. It is possible that two authors addressed the same problem, which is agreeable, but if their methodology is also the same then it amounts to plagiarism.
2. If methodology is different then the system performance also differs.
3. A research is carried out for finding an alternative hypothesis and innovative methodology that can resolve a problem effectively [2].

Therefore, goals of two research can be same, but methodology should be unique, failing which means that the hypothesis is null.

Plagiarism.org, a website dealing with the topic of plagiarism in detail, defines plagiarism as:

1. Copying content in form of words and ideas published by other authors without specifying original reference.
2. Stating or converting other person's efforts as one's own.
3. Incorrect presentation of important statements, for example, failing to insert quotation marks for attributed quotes.
4. Generating incorrect information with regard to source of quotation.

5. Modifying content by replacing words and maintaining the sentence layout without giving any reference.

As stated earlier, plagiarism is an act in which knowingly or unknowingly authors use information specified in someone else's research work. Thus, it is important to evaluate content originality before stating ownership. This is applicable to academic as well as other publishing disciplines. There has been significant growth in recent research work [3] in the field of information and technology; various algorithms have been designed and problem-solving strategies have been evaluated. Thus, it is all the more important to evaluate an author's originality rating prior to assigning any grade to his/her efforts. The foundational intent of our proposed system is to reveal the quantity of unique effort, referenced work, and online referenced or copied information in order to grade the quality of the work. This requirement directs us to the problem statement that accepts a text file, termed as document in query or question, and designs a research strategy to accomplish the task of estimating uniqueness of the document from previously available resources, both offline and online. Quality features, such as accuracy and efficiency, are the key measures of the proposed system. Our research highlights on optimum search time, scalability to acclimatize numerous mapping, and similarity algorithms. The process maps to technologies that incorporate data mining/web mining, transformation of offline dataset from existing research resources, and arranging unstructured data into training sets, using various machine learning methods for classification of content maps cluster. In this chapter, we address the text document similarity issues using cosine similarity and evaluate the efforts in the form of scalable architecture that uses recent map-reduce strategy for mapping of sentences and terms.

In order to achieve the objective of evaluating the original effort in a literature, a hybrid learning model is designed, which incorporates data mining algorithms, clusters, mappers to classify terms, sentence and documents including similarity measures to decide for exact sentence match and partial sentence match from internal and global datasets. Methods that decide the plagiarism score and rank sentences defining the plagiarized content have also been discussed in this chapter.

3.2 Literature Survey

Our literature survey aims to study the existing resources with respect to the proposed plagiarism system. We performed an exhaustive literature survey in order to understand the strategic solutions to the problems in the domain and to also identify existing potential challenges and methodologies. The outcome of the study involves discovering issues, which provide us guidelines, risk assessment, and scheduling activities in an efficient manner while designing the system.

The objectives are as follows:

- How to retrieve articles and identify and access key information and transformation?
- Which information classification strategy should be applied?
- How to choose machine learning-based algorithms that suit a specific task?
- How to identify algorithms used in text similarity, partial text mapping, and similarity score estimation?

The proposed system is scalable. Subtask division is essential for faster execution, and flexibility to adapt numerous algorithms is a challenge. Thus, in order to achieve these objectives, a few valuable resources are stated herewith. Wael H. Gomaa [4] highlighted text similarity, wherein they classify match association into three categories as corpus, knowledge, and string, based on the matching. In [4], the author compares a total of eight text similarity algorithms. The author joined multiple similar algorithms to form hybrid combinations and tested them to conclude that the best performance is achieved by a hybrid representation. Franca Debole [5] stressed on machine learning by designing a text-based classifier. In his article, he discusses how to choose a term and apply weight to the selected term. The research proposed to substitute inverse document frequency (IDF) by grouping based on term assessment function. The scope of research focused majorly on supervised learning. However, other machine learning strategies are also studied. Alexander Budanitsky's [6] research highlights on natural language processing (NLP). His article proposed a graph alike WordNet to reveal the text semantic relatedness. The point of stress in his work is that lexical semantic distance is the key in revealing semantic association. Semantic mapping outperforms distributional similarity measures for NLP-based applications. David F. Coward [7] published a book on creating dictionaries. His research emphasizes on multi-dictionary formatter. The discussion focused on folk taxonomies, classifying content by means of syntactic classes, etymologies, and identifying parts of speech. The book is a good resource for understanding the supervised learning model. Joanna Juziuk [8] discussed a catalog-based classifier. In her research, the content for categorization was derived using catalog pattern. For some patterns, a short pattern description is mapped. The article identified four dimensions of patterns for multi-agent systems, namely inspiration, abstraction, focus, and granularity. Marjan Eshaghi [9] discussed a benchmarking tactic that integrated the practice of web mining technique with standard web monitoring and assessment tools.

Literature resources aid in setting up the actions discussed in the proposed system as an algorithm assortment for data mining, designing of text classification methods, algorithm for similarity score estimation, and framework for adaptation of entire research work. Choosing a related machine learning algorithm and understanding how supervised, unsupervised, and hybrid cluster learning models can be applied is an important deciding factor. The references also incorporate supplementary sources that built up the understanding concerning the research, namely the JADE-based development [10] and the text extraction toolkit that describes the use of PDFLib [11] for PDF-to-text extraction.

We conclude that the software requirements are large, but in few months we would address the issues by developing a machine to evaluate sentence similarity between two documents. In the initial version, we use two document comparisons. Later, a multi-document comparator would be developed using clusters. We need some local training set against which the document in question will be compared. We need supervised as well as unsupervised learning algorithms that map document similarity. A hybrid model is most appropriate for such a task. Upon studying Boolean similarity, vector space similarity, and semantic relatedness similarity measures, cosine similarity is found as a good approach which can handle paraphrasing to some effective offset. Multi-agent system (MAS) framework is found to achieve the flexibility to adapt new upcoming changes in algorithm, which can be incorporated in our future work by inputting data and achieving parallel performance both as a distributed and as a web-based product.

3.3 Research Methodology

Plagiarism System sets a lot of technical issues to be considered while implementing a well-organized framework. Indeed, its learner model discovers data classes by using statistical analysis and deep mining of content to infer new knowledge that adds to precision of results and optimization of content for faster retrieval. Tuning of dependencies helps in achieving higher impact, scalable dataset, and finally improving queries hit ration. As far as the research is concerned, the subject demands intense research in areas such as text mining, document mining, and web mining. Some areas for better understanding and implementation are as follows:

- Text similarity properties, sentence similarity properties, and document similarity properties

- Machine learning, supervised and unsupervised learning methods, and hybrid approach to plagiarism detection
- Machine intelligence
- MAS framework

The research domain incorporates text mining, specifically document mining, web content or similarity mapping, sentence similarity measures, and machine learning for information classification and retrieval.

The typical research question that arises is how to reveal the strategy involved in local and web document mining, as a document is made up of text, images, charts, tables, and so on. The question emphasizes on the need of selecting a strategy to mine the document. Characteristics involved in Document Similarity utilize empirical parameters that are applicable to plagiarism domain. The next question that arises would be: what are the recent trends in machine learning which enable the selection of the most suitable algorithm for plagiarism detection? Such an algorithm is a self-learner. Third question is: what design strategy is applicable for such a domain with incremental resource, varying nature of data, and similarity properties? The solution to this should focus on identification of untouched areas, risk involved in implementation, and upcoming need.

Literature survey serves the basis for scheduling the activities involved in a system, such as algorithm selection for data mining, text classifier essential for plagiarism detection, similarity measures, and agent-based framework. Machine learning algorithms help to understand supervised, unsupervised, and hybrid cluster learning models.

Wael H. Gomaa [4] discussed text similarity by categorizing similarity into three approaches, namely corpus-based, knowledge-based, and string-based similarity. In the survey, analyses on eight algorithms were valued distinctly, and a hybrid combination of two to three algorithms were tested to its value. The best performance was achieved by hybrid model.

Franca Debole [5] emphasized on supervised learning for text-based classifier. Debole's discussions involve term selection and term weighting with supervised learning-based classifier. It was proposed to replace idf (inverse document frequency) by category-based term evaluation function. The text was limited to supervised learning and no discussion about unsupervised model was found here.

Alexander Budanitsky's discussion [6] involves a WordNet-based approach to resolve semantic relatedness. For NLP-based systems, the evaluation of lexical semantic distance is essential and further concludes that semantic relatedness outperforms distributional similarity measures for NLP-based applications.

David F. Coward's book [7] highlights his research on creating dictionaries by using Multi-Dictionary Formatter. The discussion highlights folk taxonomies, syntactic classes, loans and etymologies, and parts of speech. This approach serves as a best feed in supervised learning models.

Joanna Juziuk [8] derived the data for the classification from catalog pattern categories and short pattern description. Based on the analysis of the data, four dimensions of patterns for multi-agent systems were identified: inspiration, abstraction, focus, and granularity.

Marjan Eshaghi [9] deals with a benchmarking methodology based on the integrated usage of web mining techniques and standard web monitoring and assessment tools.

The references even include added sources that built up the knowledge on the subject, such as JADE-based development [10] and text extraction toolkit that describes the use of PDFLib [3] for PDF-to-text extraction.

The following issues in the existing methods were revealed after research:

1. Plagiarism is a global issue where copyright is applicable to text, graphics, sound, stories, and so on.
2. Multilingual plagiarism detection is still a challenge.
3. Datasets for online web page plagiarism are limited to reachability of search engine crawlers and access protocols.
4. Exact matching of text similarity is possible, but finding a method to minimize the number of comparisons and to decide plagiarism is still a challenge.
5. The minimum number of comparisons needed to answer the worst case that contents are not plagiarized needs to be fixed.
6. Deciding parameters that formulate the results for plagiarism system.
7. The strategy to identify the exact documents that are likely to be similar and avoid checking unnecessary documents.
8. Self plagiarism is also a challenge.

3.4 System Architecture

Client–server architecture with the server stores the training set, client requests for effort estimation score and uploads content to server. Care is also taken that the software estimates the result online as well as offline.

The system involves text-based plagiarism detection. It involves data mining, document retrieval, extraction, machine learning, mappers, clusters, distributed data access, web information access, text similarity measures, sentence similarity measures, and hybrid architecture that is flexible

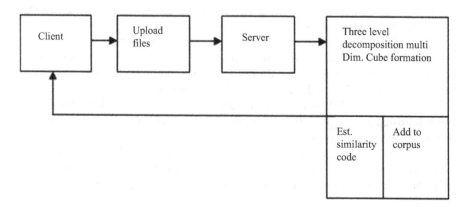

FIGURE 3.1
Client–Server module for plagiarism evaluation

to adapt innovative algorithms and improve system accuracy and makes the system scalable. The system can be scaled after involving various plagiarism detectors such as UML Diagram Checker, Plagiarized Images Checker, Duplicate Chart Detector, OCR Mapper, Multilingual Translator Mapper, and so on.

Software-based effort detectors have drawbacks because of lack of human-like intelligence to sense any associated terms. Semantic related-ness detection are complex, multilingual issues where documents are translated as it is from original source. Complex computations involved in understanding the framework is one of the few challenges in research. Thus, an ongoing research is essential to evaluate such applications.

Human intervention might decide on plagiarism, but computation can outperform in terms of detecting faster than humans.

Data collection for implementation purpose is achieved by analyzing the following sources of information:

1. Publication or institute-owned database (internal dataset).
2. Open-source information published on web (global dataset).
3. Other restricted online resources accessed by using authorized account of author.

Here the system implementation is achieved by checking internal and global dataset to evaluate plagiarism score. Point 3 is dependent on API interfaces that link to such private publication databases, which are currently not implemented in our work.

After studying the huge number of articles that are available on text plagiarism with regard to existing systems, the software are able to compare two files at a time, or some text-based comparison limit, to detect modification of sentences. Thus, their precision is limited to the algorithm used. It shows that the hybrid approach is best considering the current problems, hence in our implementation a hybrid model that uses multilayer and multi mappers are used. Hybridization incorporates both supervised and unsupervised learning and is made extendable to adopt other learning aspects such as temporal learning. Clusters have a remarkable performance as far as auto-classification is concerned. Selection of efficient searching algorithms based on algorithm searching complexity studies specify that array-based word comparison has worst case complexity of $O(n)$, binary Tree with $\text{Log}(n)$, Hashing $O(1)$ to $O(n)$. However, indexing provides key-based search, which is quite suitable in the plagiarism checker system. Thus, here hashing as well as indexing are used as a combined approach. During the study, we also found out that the text mining domain is essential for research being carried out. Text mining is an applied or essential method by which text-based analysis can be performed. There are various modes in which text can be mined. A variety of applications are involved based on text mining. Our discussion here is to design the best mining technique for plagiarism detection.

Analysis of documents: Based on the above stated approaches we can analyze a text document based on document relevance or words-based correlation. The core of plagiarism detection lies in content analysis [12], that is, mapping documents whose content is similar to the document in question. The content here is text, hence the text mining approach is applied. The analysis reveals that text document is a collection of sentences, whereas a sentence is a collection of words. Thus, a mapper is introduced to map sentences to document and map words to sentences. The mapper contains link information on the files that are likely to be containing sentence plagiarism. The term mapper maps the sentence beginning from where the term exists; its significance lies in detecting semantic relation and partial match detection.

System complexity: The system accepts a text-based input by means of text file *.txt, MS Word content, or text extracted from *.pdf files. Plagiarism system involves sentence comparison of input files with other files in dataset or uniform resource locators (URLs) from the web. The function of similarity association is mapped by using indexing technique. Information similarity mapping with input files is described subsequently.

Consider only "Sn" lines or sentences. Each file in the dataset is assumed to contain "Sx" sentences, and the dataset contains "Fx" files, each sentence contains approximately "w" words. If Sn = 10, Sx = 10, Fx = 2000 files, w = 10, then in order to obtain the results in worst case, sentence mapping requires the following comparisons and Input/Output overhead:

$$\text{Total number of files} = 2000$$

Total number of sentences St = Fx × Sx = 20000 sentences

Number of comparisons required to detect if a sentence Si is plagiarized is 20000.

Therefore, to conclude that the sentence is plagiarized or not we require Sn × St = 10 × 20000 = 200000 comparisons, and this complexity increases with the size of the dataset. Hence the system performance degrades. We applied the method of reduction that reduces the number of comparisons by using key-based mapping and indexing. However, this key-based approach has a disadvantage, that is, it is best for exact sentence match, but a single modification in input sentence can pass the plagiarism detectors. In order to remove this defect and improve accuracy, our proposed system uses dual key, a sentence key, and word- or term-based key, i.e. TermKey. But TermKey involves increased number of comparisons.

Research implementation strategy—mappers and reducers: An association model such as key, value pairs is stated as mapper (k,v) where mapper is a set of the pair k→key, v→value Thus in implementation two keys are produced one for sentence mapping and another for term to sentence mapping.

Sentences having the same key are clustered into sentence cluster, whereas the terms having similar key are clustered into term clusters.

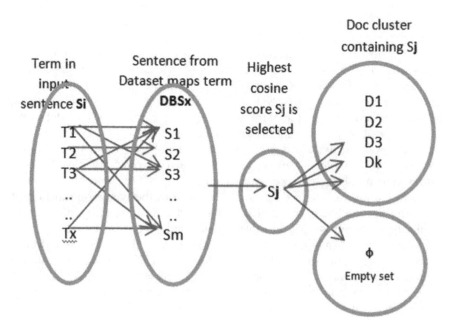

FIGURE 3.2
Term-to-sentence and sentence-to-file mapping

Thus, two types of clusters are mapped by the mappers. Figure 3.2 depicts term-to-sentence and sentence-to-file mapping.

Sentence mappers: Every file is a collection of sentences, whereas each sentence is a collection of terms, stop words, and stemming. Supervised learning is applied to detect sentences based on rules of grammar that define syntax to extract words from files in question. Pointing is mapping the key/value pairs where each key is assigned a location vector as filename, file path, and line number where the sentence exists in dataset files.

Term mappers: Term mappers add accuracy to the system where the TermKey is assigned a value sentence key, that is, term exists in some sentences and those sentences are located in some file.

3.4.1 Proposed Architecture

The following software architecture highlights the core components involved in each activity of development. Our architecture ensembles multi-layer approach to plagiarism estimation as seen in the architecture given in Figure 3.3. Each layer is assigned a task that can be executed in parallel. A multi-agent-based framework is adaptable to such a design.

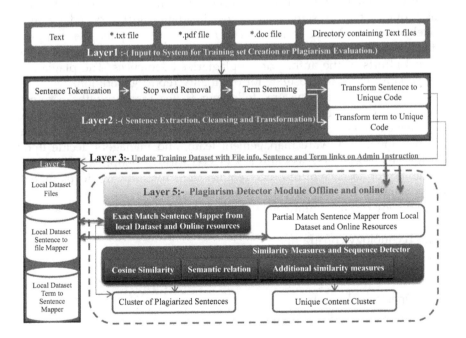

FIGURE 3.3
Layered architecture

Layer 1 handles the task of text-mining, document retrieval, and conversion.

Layer 2 processes the data for cleansing and transformation in a structured way. It converts the unstructured data into a structured form so that it can be easily analyzed. In this layer, the documents are tokenized based on sentences to produce a sentence vector. Each sentence from the sentence vector is cleansed by removing stop words and stemming so the number of comparisons will be reduced and some stemming errors in comparison can be minimized. Furthermore, each sentence is tokenized to terms and a term vector per sentence is obtained. Each sentence and terms are converted to unique codes for accurate identification. Most of the data mining and supervised learning mechanisms are applied here for local dataset creation.

Layer 3 dispatches the outcome of Layer 2 to either updating of local dataset or to plagiarism detector. The core functionality is to provide dataset connectivity to save new records or pass records to the next layer for further analysis.

Layer 4 is a simple local data store that contains files or documents, sentence-to-file mappers, and term-to-sentence mappers.

Layer 5 is the Core Plagiarism Detector Module. It involves the process for local dataset mapper and online resources mapper. It has a similarity estimator layer that adapts various similarity measures; we use the cosine similarity evaluator for sentence mapping. It is the most important layer that provides two clusters as outcome. It adapts unsupervised learning model for cluster-based learning. The first cluster contains the mapping details of plagiarized sentences, while the second cluster dispatches the original efforts of the author or un-plagiarized sentences. The plagiarism score finally serves the baseline for originality report generation.

While designing the system, following issues regarding the resource possibilities or feasibilities such as technical constraints, time constraints and financial constraints or budgetary constraints need to be addressed. The developers, software management team, and managers decide upon whether to accept or reject the job of software development, based on availability or absence of resource requirements for software development, and constraints that are involved in the life cycle.

We hereby discuss following six limitations of this study:

1. Technology constraints
2. Time constraints
3. Economic or financial constraints
4. Operational constraints
5. Environmental constraints
6. Alternative solutions

After completion of the feasibility study, following technology constraints were observed. Technical tools and readiness of other technical resources are main concerns. There are technical domain issues as well. A questionnaire was prepared to answer technical constraints, which is as follows:

i. Development platform: Red Hat Linux

ii. Software development language and tools availability

iii. Quality attributes involved in the system

iv. The feasibility of proposed algorithm or methodology

v. Deployment environment as web, distributed, or virtualization as cloud

vi. Are the product requirements genuine and implementable?

With regard to operational feasibility, the product will be useful to detect plagiarism; it has a potential demand in the field of literature publication, academic effort evaluation in various universities and also in conferences arranged by organizations over the world. Hence, the product has a scope in the business world too [2].

With regard to time constraints, the system scope is not limited to a particular algorithm implementation, but it involves improvement over issues involved in text comparison measures. Thus, subsequent versions of the software may be needed; it is a scalable software product. Each version will add more functionality and accuracy to the system, and thereby increase the software quality and capability.

The substitutable or alternative solution might be an alternative implementation through sentence-based cluster mapping, which could be faster but with lesser accuracy. Another implementation would be the re-enforcement learning based on feedback sessions, which could improve accuracy but the system would lack the self-learning capability, which indirectly is supervised learning. Thus, a hybrid model with sentence and term clusters are developed in our system with a reduction mapper to evaluate accurate results. The mixed model implemented here involves both supervised learning and unsupervised learning, thus integrating a self-learning model expected from an intelligent machine. Even though it is a self-learning system, it lacks in decision making as human beings, and thus requiring a reviewer to analyze the acceptance of results because sometimes paraphrasing detected by software might be allowed by a human reviewer.

Details of Plagiasil System algorithm: We will elaborate the implementation goals and objectives of the system here.

- Implementation goal is to ensemble a software system that is adaptable or flexible to accept changes and to integrate various related components to develop a working model for effort evaluation of the literature to be published or academic submission to be evaluated.

- The purpose of the software-based evaluation is to save time, gain efficiency, accuracy and support in decision making by evaluators in order to accept resources or suggest some modifications.

3.4.2 Discussion on Problem Statement

The input document that would undergo testing can be in the form of plain text, open-xml format (MS Word document), portable document format, and so on. We design a software process for evaluating the efforts of the author of the document in producing that content (resource). Resource mapping must involve a comparison based on local database and should also mine web-based literature to combine a detailed content originality report.

The challenges in system implementation are as follows:

- There are data mining issues in comparing amorphous data with multiple documents.
- Comparison dataset grows as new resources are added to the local dataset; system performance may degrade as the number of comparisons rise.
- Web mining strategy is applied for faster estimation of efforts in analyzing global published data.
- Studies involve content extraction and conversion from various text formats (file layout transformation to text).
- Algorithms for reduction of comparison, adaptation of software patterns to improve accuracy, and integrated components are necessary.
- Design architecture to adapt multi-layer similarity methods for sentence comparisons are required.

Product Core Algorithm for Plagiarism Detection: Supervised learning provides the core functionality to recognize tokens and stop words. Hence, rules can be defined to remove stop words from the sentences. Cluster-based mappers improve similarity detection and reduce the number of comparisons involved in unsupervised learning. Therefore, we conclude that a hybrid learning machine suits our requirement the best. There are two algorithms involved in our system: server algorithm and client algorithm.

3.4.3 Server Algorithm for Training Dataset Creation

1. Read file
2. Break file into sentences; output list of sentence.
3. Remove stop words

4. Remove stemming words
5. Generate Sentence Unique Code (SUC) for each sentence. S_Hash-code.txt {filename: line_no}
6. Update table with fields: T1{SUC,Value}
7. Update Table: T2{SUC,WUC,W_value}
    a. For each term in sentence, say s1=w[n], i.e., sentence s1 contains n terms
    b. We find such s1, s2, s3, ..., sm is an array of sentence in a file.
    c. Query select unique(SUC) from T2, where w_value like('term1') or w_value like('term2') or w_value like(term N)...
    d. End of Training set preparation.
8. If all files are covered then stop, or else go to step 1.

This algorithm is a hybrid learning model that comprises supervised as well as unsupervised learning.

3.4.4 Client Algorithm for Amount of Plagiarism

1. Read file
2. Break file into sentences output, Array of sentence, say Sx[].
3. Remove stop words
4. Remove stemming words
5. Generate SUC code for each sentence and Sj array of sentence in Query file.
6. Each sentence contains Sj; it contains array of terms, say, Ti terms for that sentence.
7. For each term Ti in sentence Sj, fire the query as select unique(SUC) from T2 where w_value like('term1') or w_value like('term2') or w_value like(term N)...
8. Output list of sentence that contain the terms.
9. Calculate cosine transform for each Sj with SUC:file:line_no ouput score [0–1] since $\cos(0) \to 1$, i.e., 100% match anything nearer to 1 is accepted as plagiarized else not.
10. Count number of plagiarized sentences and calculate similarity score.
11. Generate HTML/PDF report

3.4.5 Software Design

Functional requirements: The core system functionalities that are required by domain users, end users, and the administrator are covered here. It describes the features that are required for the system to fulfil user demands, some of which are User Interface, reports, meta-data information, and database required. It adds to the usability and reliability of software product.

Non-functional requirements: These define the limitations, validations, and rule-based implementations. These non-functional requirements improve the system quality. Their existence in the system improves system efficiency, and they also act as supporting agents to guide user to reach the expected objective. Login name, password, response time, request time, deciding auto backup, auto recovery, authorization services, data field validation, reduction of SQL-Injection attack, avoidance and prevention of data in transmission, storage and maintaining service logs are the few non-functional requirements that are constraints.

Product perception: We believe that unsupervised multi-layer Plagiarism System is an efficient system that adopts a hybrid approach to improve system accuracy and performance. This product is essential in the upcoming times in the fields of literature publication and academic author efforts estimation. It has an enduring scope in business and research as well, as these fields too need measures to improve plagiarism detection.

Design and implementation constraints: This product is developed on J2EE platform in order to attain portability and web-based access. Even though the product design is flexible, it is limited to desktop-based usage. An Internet connection enhances the system accuracy. The current system has not considered mobile-based services. It has been developed for academic purposes and in future would need refinement on commercial aspects. The system accuracy can be improved by adding rules. At present the file access is for text-based assessment, which needs to be elaborated for inclusion of other formats such as images and audio charts.

3.4.6 System Features

The software has a generic login and user registration module, which formulate the activity area and apply constraints on system access. The administrator is the only controller of user and system management. The administrator can upload a file, check for plagiarism score, and update the training dataset. Furthermore, the administrator is given the flexibility to add stop words, word stemming rules, and so on. The user of the product can register and request for a service. Registered users can access the system by uploading the text or entering the text into the submission form. A plagiarism score with a detailed report is produced in the form of an HTML/PDF document.

The server acts as a central processing element that stores the uploaded files. These uploaded resources are from publication literature, reports from students or research scholars, online resources, and so on. The server pre-processes these files by applying data decomposition algorithms and maintains the classified contents to help faster retrieval. Data decomposition classifies the content easily and directs search execution. The server plays a major role in evaluating the author's efforts behind the file in question. In order to uncover the plagiarism score, the server maintains the training set that comprises of the above-specified resources. Details regarding creation of the training set and data decomposition are further revealed herewith.

Data decomposition by a set of constraints and fine-grained analysis: The research input constitutes of text files—a research paper is preferable. Journal papers are divided into three base classes. Some of them are from references sections, comprising sentences that are properly citied, and others are the original efforts of the author. Thus, the system ensures that the data classes as referenced work, plagiarized content, and non-plagiarized content. Hence, the decomposition algorithm first classifies the files into sentences and references. The sentences are further classified into cited content and non-cited content. These contribute to the fine-grained analysis by further decomposing the sentences to words or valuable terms. The fine-grained content is best suited for parallel applications that speed up query processing and give quick response time.

We define various levels of decompositions in the following sections.

3.5 Level 1 Decomposition

The plagiarism system recognizes participation as a file or folder to create dataset that serves as training corpus.

$$f_i = \{\text{file}\#1, \text{file}\#2, \text{file}\#3, \ldots\ldots\ldots \text{file}\#n\}$$

f_i is a set of files in the corpus, and **file#i** means ith file in the corpus. In other word file#i acts as a handler to access the resources.

Where i= 1, 2, 3, ...n

The training set to the system is scalable and new contents are validated as genuine contents, or say plagiarism free contents. Datasets are dynamic in nature and tend to grow with age. This dynamic updating needs a scalable data management process. Content organization algorithms will update the entire structure as new resources are added to the corpus. Another issue handled is the reading of multi-formatted documents. The input resources accepted by the proposed system are text files, pdf files, word documents etc. Files such as latex, html, and other textual contents scales the digging of resources and adds capability to the training corpus.

3.6 Level 2 Decomposition

Medium-grained decomposition of file into sentences classifies the content into three stated classes: referenced sentences, non-reference sentences, and citations.

Each file is further segregated into sentences.

$$S_j = \{\text{Sent\#1}, \text{Sent\#2}, \text{Sent\#3}, \ldots, \text{Sent\#}n\}$$

where S_j is a set of sentences in the file, and Sent#j means j-th sentence in a file. "n" terms to number of sentences in any file#i.

where $j = 1, 2, 3, \ldots, n$

Classification table

Reference sentence	Non-reference sentence	Citations/references
Sent#5	Sent#12	Sent#100
Sent#7	Sent#15	Sent#101
Sent#11	Sent#18	Sent#102
...	...	...
...	...	...

Significance of such classification is in eliminating content and deciding to map as plagiarized or not.

3.7 Level 3 Decomposition

This level deals with the formulation of cluster nodes for author effort estimation by using vectorized non-referenced sentences. Such sentences are estimated for classification under genuine work or copied content either as a part or whole. Each sentence comprises of words, with say an upper bound of 10 words in a sentence that needs to be segmented.

Data mining of fine grain contents is uncovered here as:

$$\text{Word\#}k = \{\text{Word\#1}, \text{Word\#2}, \text{Word\#3}, \ldots \text{Word\#}n\}$$

where $k = 1, 2, 3, \ldots, n$

3.8 Noise Reduction

This step estimates the non-significant words from the sentence; it is a form of optimizing the training set. Content that just add to the meaning of the sentence but the actual presence or absence of such content does not affect the meaning of the sentence are considered insignificant. Removal of such terms reduces the dataset size, which leads to a smaller number of comparisons and faster results which are indeed the goals to achieve.

Supervised machine learning approach is best suited to implement a mapping of input to output pair. The input Sent#k vector is further decomposed into two classes, namely valuable words and non-valuable words or irrelevant contents. For example, the words such as **"the, for, a, an etc"** are rated as non-valuable words as their meaning does not highly impact the results. Classification algorithms play an effective role in designing such eliminating terms. Since these terms are of shorter lengths, they can be loaded in RAM while processing. The algorithm is static in nature and must be balanced based on the training set. Frequency of access of such terms is very high, and search can be prioritized based on the hit count of the content.

The real skill is to develop a machine learning strategy for unsupervised contents by auto-classification of fine-grained contents. This classification signifies extraction of complex associations, meta-info, semantically mapping to existing terms based on system knowledge, and finally handling uncertainty in data. Organizing such content in the form of WordNet Classification and Regression Tree (CART), lattice organization is normally expected but the limitation of the RAM size hinders the compensation of the corpus. Since the training contents are scalable, such corpus may lead to content replication, nodes repeating in graph, or visiting nodes repeatedly during traversal of nodes. Also, the system demands semantic analysis to show a more precise result.

Transformation of training set: The fine-grained contents are further packed to multi-dimensional Data cubes. Each file is transformed to such cubes. Figure 3.4 shows a sample architecture for cube organization of decomposed data. This formulates a file node and serves as a training set.

Grain packing to node where each weighted word is provided with unique identifiers is reflected in the words template above. Word vector is packed with a unique identifier which is a subset of each sentence. The sentence vectors are packed further in to nodes of a file. Finally, the file vector comprises the content of the corpus. File handlers are unique identifiers to access the resources easily.

3.9 Client-Side Query Processing Sequence

The process is simple enough as shown in the Figure 3.4. Similarity score estimation process defines various levels starting from coarse-grained analysis to fine-grained analysis as follows:

a) *If two nodes matching two files are similar*:
   Boolean fileMap(File f1, File f2) (i)
   It maps to the reduced score of two files f1 and f2. Reduce score is nothing but a compressed value that can be compared easily. It implies that fileMap() just maps the compressed value to two file nodes to reveal an exact match between two files.

b) *If sentence vector from node maps to similarity then the sentence is plagiarized*:
   Boolean SentMap(Sent S1, Sent S2) (ii)
   Direct sentence-to-sentence map uses either text matching tools or compressed sentence mapping. Compressed sentence mapping generates the overhead of compression and decompression, but the overall performance is improved when larger number of comparisons exists.

c) *Partial sentence mapping is achieved using cosine score of the sentence vectors*:
   float CosineScore(Sent s1, Sent s2) (iii)

d) *Cosine score estimation*: It accepts two vectors s1 and s2 and returns a matching score between [0,1] where zero terms exacts match and value 'one' indicate non-matching vectors.

e) *Semantic sentence mapping*: It is mapped by applying the unique key identifier to similar meaning words as shown in Figure 3.4.

f) *Total plagiarism Score (Tps)*: It is mapped by applying the total score estimation which is achieved by scanning each of the non-referenced contents and a cumulative score for the file in question is evaluated.

$$\text{Tps} = \sum_{i=0}^{n} \text{matchScore}(\text{file fi, dataSet node}) --(v)$$

where fi is the file in question decomposed to three levels of fine-grained vectors to map similarity score with training corpus residing on the server.

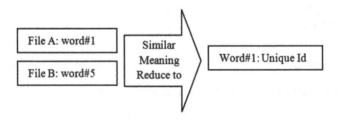

FIGURE 3.4
Semantic words mapping to unique identifier

3.10 Cosine Score Estimation for Partial Mapping or Partly Modified Sentence

The most complicated decision maker of the system is the cosine score. It takes the decision of partial matching vectors. The strategy is to create two sets that may not be ordered. The length of each term in the vectors is estimated. The contents are mapped first by length and then by computing the dot product as described in equation (iv).

(i) Set s1 and s2 as two sentence vectors.

(ii) Estimate the length of each element in the two sets and compute the **dot** product of the two vectors s1 and s2

(iii) The cosine angle can be computed further as:

$$\mathrm{Cos}(\theta) = (s1 \cdot s2)/(|s1| * |s2|) \tag{iv}$$

(iv) Result $\cos(q) = 1$ terms exact match of competent vector, any value nearer to 1 implies partial map.

(v) An offset for partial match is set to 0.6, else the decision can be made that the two vectors differ. This offset can be set based on percentage of allowed similarity in contents.

3.10.1 Software System Performance Evaluation and Results

The software system is a three-tier system and possesses multi-module functionality. It comprises of an administrative module and an end user module.

Administrative module for plagiarism dataset generation: This module's core activity is to facilitate a training dataset or supervised learner module. It includes the following functionalities:

- Upload file for generating a local dataset
- Upload files to test for software functionality
- Configure the stop word rules
- Configure stemming criteria
- Configure result offset
- Block or enable/disable user profiles and ACL List

The end user module for plagiarism evaluation: Its core activity is to generate content originality report for the user requested file.

- Upload files for evaluating author efforts

- Offline dataset mapper
- Online dataset mapper
- Generate plagiarism or originality report
- Request for service update

3.10.2 Plagiarism System Performance Enhancement

During information retrieval, the core is to extract the most relevant document in order to detect the most relevant sentence that maps the sentence in question. The sentence that is detected as twin in the plagiarism system might be an exact match or a partial match [13, 14]. In case of modified sentence and paraphrasing, all or some of the terms in the sentence are altered. Detecting such sentences might introduce error thereby affecting the accuracy. The resulting evaluating parameters can be classified as follows [15].

(i) Identification of true positive (TP) results: Sentences that reflect the exact match or reveal to have similar meaning, which relates to the concept is termed as TP. In plagiarism evaluation, the sentences that map to TP are evaluated as plagiarized.

(ii) Identification of true negative (TN) result: Set of sentences that reflects as unique and do not map to any data set contents are mapped as true negative. Identification of error in set of false positive (FP) and false negative (FN) sentences is challenging and raises confusion in verifying the result to check whether the sentence is replicated from some source or it is author's effort. If the percentage of FP and FN set sentences are more, then the accuracy of the system is questionable and the algorithm needs to be redeveloped or similarity measures revaluated.

Some measures that quantify the Plagiarism System performance for effective retrieval are as follows [15]:

(i) Precision for plagiarism system: It is defined as a proportion of the reclaimed sentence that is retrieved by the system in order to declare the similarity to the sentence in question. As we can match sentences to evaluate plagiarism score, our application precision will be based on the following equation. The significance of precision is that it computes the percentile of positive estimates that are precise.Precision estimate takes all sentences into account, but it can also be assessed at a certain predefined cut-off threshold, considering only the upper-ranked results responded by the system. Such estimate is termed as precision at n or P@n.

$$\text{Precision} = \frac{|\{\text{Relevant Sentence}\}| \cap |\{\text{Retrieved Sentence}\}|}{|\{\text{Retrieved Sentence}\}|}$$

As an example, case for a sentence hunt on a set of sentences precision is the amount of accurate results divided by the amount of all retrieved sentences in results.

(ii) Recall for plagiarism system: It is a proportion of the sentences that is relevant to the sentences in question that are effectively fetched or retrieved. For our plagiarism, similarity evaluation recall stands as follows [15]:

$$\text{Recall} = \frac{|\{\text{Relevant Sentence}\}| \cap |\{\text{Retrieved Sentence}\}|}{|\{\text{Relevant Sentence}\}|}$$

Significance of recall is that for a sentence evolution on a set of sentences, recall is the sum of precise outcomes divided by the amount of sentences from results that are supposed to be retrieved. Recall is also termed as sensitivity or positive predictive value.

3.10.3 Performance Parameters

Performance measures are highlighted in Figure 3.5 with relevance to results.

As per our plagiarism evaluation module, the sentences in question are classified in to two clusters: one cluster of plagiarized sentences and another cluster of unique sentences that reflects owner's efforts. The

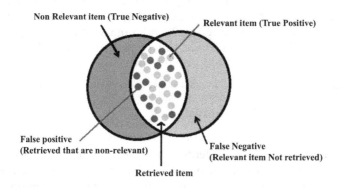

FIGURE 3.5
Performance measures

clusters might be misclassified if it contains a set of FPs and FNs and such conditions give rise to confusion as shown in Figure 3.5.

The accuracy of classification is given by

Accuracy = (True Positive + True Negative)/(P + N)

where P and N are positive and negative instances of some condition [15].

Sample calculations that reflects results: We predict some data for the demonstration of evaluation of precision and recall as follows: Input a document in question for plagiarism score evaluation containing 100 plagiarized sentences among 10,000 sentences. The issue is to predict which ones are positive. The Plagiasil system retrieves a score of 200 to represent the probability of plagiarism, w.r.t. 100 positive cases. The results are evaluated several times to predict the accuracy of the retrieved sentences.

For instance, let expected results be 100 sentences.

TP = 60 sentences (found as plagiarized sentence, i.e. accurate detection)

FN = 40

TN = 9760

FP = 140

Therefore, Recall will be 60 out of 100 = 60%

Retrieved Results: **200 sentences.**

Therefore, precision counts will be 60 out of 200 = 30%

Total statements to be compared = 10000.

Accuracy of the system will be 9760 + 60 = 9820 out of 10000 = 98.20%

3.11 Experimental Results

In Figure 3.6, plots on the x-axis represent the similarity scores with groups of sentences. Here, 10 sentences are grouped. The above result estimates a file of 150 sentences and hence contains 15 groups. Figure 3.6 highlights a matching score of 10% in group1, 20% in group2, and a maximum of 90% plagiarized contents are revealed in group 10. Overall, plagiarism score estimates to 20%.

3.12 System Capability Analysis

System featured 100% accuracy can be estimated for exact matching statements. Partially modified content is revealed using the cosine score. Deeper mining improves accuracy and performance. Unique identifiers of words make it possible to map two file sets having common scores between them. Multi-document mapping is achieved by mapping the corpus on file-to-file-based content similarity analysis.

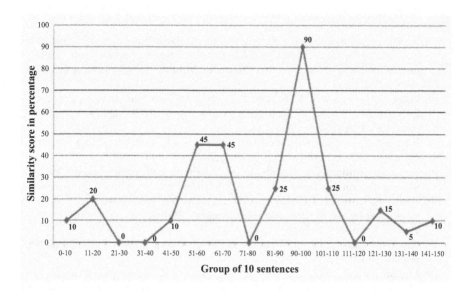

FIGURE 3.6
Plagiarism score of a document with 150 sentences with 20% similarity detection.

3.13 Conclusion

It is essential to evaluate efforts of authors in academic and research publications to analyze the quality of work and unique valuable information resource. This is achieved by means of a plagiarism detector system that applies a reduction algorithm to the hybrid machine learner model that evaluates text similarity on the web as well as in local dataset. About the future works in the plagiarism detection system, text-based replication detection is the initial factor that needs to be discussed. Other features that can be copied from literature are images, charts, templates, and so on. The upcoming research topic would be analysis of documents translated in other languages.

References

[1] David J. C. Mackay, "Information Theory, Inference and Learning Algorithms", The Press Syndicate of the University of Cambridge, Cambridge University Press, 2003, ISBN 0 521 64298 1.
[2] Kaveh Bakhtiyari, Hadi Salehi, Mohamed Amin Embi, et al. "Ethical and Unethical Methods of Plagiarism Prevention in Academic Writing",

International Education Studies Volume 7, Issue 7, 2014 ISSN 1913-9020 E-ISSN 1913-9039.

[3] Shreya P. Amilkanthwar, Poonam N. Railkar, Parikshit N. Mahalle, "Internet-SensorInformation Mining Using Machine Learning Approach," International Journal of Computer Science and Network, Volume 4, Issue 6, December 2015, ISSN - 2277:5420.

[4] Wael H. Gomaa, Aly A. Fahmy, "A Survey of Text Similarity Approaches", International Journal of Computer Applications (0975 – 8887), Volume 68, Issue 13, April 2013.

[5] Franca Debole, Fabrizio Sebastiani, "Supervised Term Weighting for Automated Text Categorization", SAC2003, Melbourne, FL, USA

[6] Alexander Budanitsky, Graeme Hirst, "Evaluating WordNet-based Measures of Lexical Semantic Relatedness," Association for Computational Linguistics.

[7] David F. Coward, Charles E. Grimes, "Making Dictionaries: A Guide to Lexicography and the Multi-Dictionary Formatter," SIL International Waxhaw, North Carolina

[8] "Joanna Juziuk", "Design Patterns for Multi-Agent Systems", Linnaeus University, June 2012.

[9] Marjan Eshaghi, S.Z. Gawali, "Investigation and Analysis of Current Web Mining Techniques as well as Frameworks", International Journal of Recent Technology and Engineering (IJRTE), Volume 2, Issue 1, March 2013, ISSN: 2277-3878

[10] Fabio Bellifemine, Giovanni Caire, Agostino Poggi, Giovanni Rimassa, "JADE: A Software Framework for Developing Multi-Agent Applications. Lessons learned", Elsevier, 2007.

[11] A. S. Bin-Habtoor, M. A. Zaher, "A Survey on Plagiarism Detection Systems", International Journal of Computer Theory and Engineering Volume 4, Issue 2, April 2012.

[12] Debotosh Bhattacharjee, Sandipan Dutta, "Plagiarism Detection by Identifying the Equations" Published by Elsevier Ltd, 2013.

[13] Salha M. Alzahrani, Naomie Salim, Ajith Abraham, et al., "Understanding Plagiarism Linguistic Patterns, Textual Features, and Detection Methods," IEEE, 2011, 1094-6977

[14] Zhenzhou Tian, Qinghua Zheng, Ting Liu, Ming Fan, "DKISB: Dynamic Key Instruction Sequence Birthmark for Software Plagiarism Detection", IEEE, 2013, 978-0-7695-5088-6/13.

[15] A. R. Naik, S. K. Pathan, "Indian Monsoon Rainfall Classification and Prediction Using Robust Back Propagation Neural Network, International Journal of Emerging Technology and Advanced Engineering, Volume 3, Issue 11, November 2013.

[16] Yuan Luo, Kecheng Liu, Darryl N. Davis, "A Multi-Agent Decision Support System for Stock Trading," IEEE, 2002, 0890-8044/01/$17.00 ©

[17] Biao Qin, Tuni Xia, Fang Li, "DTU: A Decision Tree for Uncertain Data", Springer-Verlag Berlin Heidelberg, 2009.

[18] David Hand, Heikki Mannila, Padhraic Smyth, "Principles of Data mining," Massachusetts Institute of Technology, 2001, ISBN: 0-262-08290-X.

[19] Eugene Santos, Jr., Eunice E. Santos, Hien Nguyen, Long Pan, John Korah, "A Large-Scale Distributed Framework for Information Retrieval in Large Dynamic Search Spaces."

[20] M. Potthast, L.A. Barrón Cedeno, B. Stein, P. Rosso, "Cross-Language Plagiarism Detection." Language Resources and Evaluation, Volume 45, Issue 1, pp. 45–62, 2011. doi:10.1007/s10579-009-9114-z.

[21] Selwyn Piramuthu, "Evaluating Feature Selection Methods for Learning in Data Mining Applications."

[22] Romain Brixtel, Mathieu Fontaine, Boris Lesner, et al., "Language-Independent Clone Detection Applied to Plagiarism Detection", Conference: Source Code Analysis and Manipulation (SCAM), GREYC-CNRS, University of Caen Basse-Normandie, Caen, France. DOI: 10.1109/SCAM.2010.19

[23] B. Kitchenham, "Guidelines for Performing Systematic Literature Review in Software Engineering," EBSE Technical Report, Keele University, Version 2.3, 2007.

[24] Swati Patil, Prof. S.K. Pathan, "Fault Prediction based on Neuro-Fuzzy System," International Journal of Advanced Research in Computer Engineering & Technology, IJERT, Volume 2015, Issue 4, pp. 514–517, June 2015. 06, 2278-0181.

Section II

Machine Learning in Data Mining

4

Digital Image Processing Using Wavelets

Basic Principles and Application

Luminița Moraru
Faculty of Sciences and Environment, Modelling & Simulation Laboratory, Dunarea de Jos University of Galati, Romania

Simona Moldovanu
Faculty of Sciences and Environment, Modelling & Simulation Laboratory, Dunarea de Jos University of Galati, Romania
Department of Computer Science and Engineering, Electrical and Electronics Engineering, Faculty of Control Systems, Computers, Dunarea de Jos University of Galati, Romania

Salam Khan
Department of Physics, Chemistry and Mathematics, Alabama A&M University, Normal AL-35762, USA

Anjan Biswas
Department of Physics, Chemistry and Mathematics, Alabama A&M University, Normal AL-35762, USA
Department of Mathematics, King Abdulaziz University, Jeddah 21589, Saudi Arabia
Department of Mathematics and Statistics, Tshwane University of Technology, Pretoria 0008, South Africa

CONTENTS

4.1 Introduction

Recent pathology practices require huge labor and time-intensive processes. Currently, advantages of digital image processing and analysis are recognizable in all areas, including pathology.

Pathologists and associated scientific teams use, among other investigative techniques, microscopic evaluation. Automatic imaging techniques provide huge advantages, including both image processing and image analysis. Digital image processing enables enhanced and virtually noise-free images. Microscopy digital image-processing techniques are used for both enhancement and manipulation of raw images to make the informational content available. Microscopy digital image analysis techniques comprehend provided data by mean of data recognition techniques, image-quality assessment, or classification. There are many image-processing algorithms and analysis techniques, and this chapter intends to underline the principles behind image manipulation.

The technique utilized by optical microscopy is transmitted light for image recording. Generally, microscopy volumes are strongly anisotropic, and acquired digital images suffer from intensity inhomogeneity, uneven illumination and backgrounds, noise, poor contrast, aberration artifacts, out-of-focus area, color shifts, and color balance errors. These artifacts lead to a bias in the results and are a cause of inaccuracy and errors. Image processing represents a variety of methods to manipulate and generate intensity changes useful for observation and imaging. Methods that manipulate the common microscope illumination modes are brightfield, phase contrast and/or differential interference contrast, and oblique. Other methods are specific to the MATLAB® programming language, namely, the Wavelets and Image Processing Toolbox for MATLAB R2014a software.

The methods under consideration belong to unsupervised machine-learning methods. We proposed some algorithms for wavelet transforms to obtain texture information and a machine-learning approach to handle the de-noising operations that become more adaptable when the wavelet transforms are integrated in decision (Suraj et al., 2014; Moraru et al., 2017; Rajinikanth et al., 2018). To enrich our study of edge detection, we included two machine-learning techniques: image enhancement by wavelet transforms and image fusion. This chapter presents the fundamental tools of digital image processing based on wavelets applications. The principles of wavelet representation, image enhancement such as filtering for noise reduction or removal, edge detection, edge enhancement, and image fusion are discussed and some applications are presented. At this point, edge detection and fusion technique related to horizontal and vertical orientations are highlighted. A three-step framework is exploited: decomposition, recomposition, and fusion. The image-quality assessment is discussed in terms of objective metrics such as peak signal-to-noise ratio (PSNR) and mean square error (MSE) and of a hybrid approach such as structural similarity (SSIM) index as a method to establish the similarity between two images. For edge detection, the latter method takes advantage of sensitivity of human visual system.

In this study, the problems and procedures are presented in two-dimensional (2D) real images, which are available for free access in the Olympus Microscopy Resource Center galleries (http://www.olympusmicro.com/galleries/index.html). The same dataset of test image is used across this work for comparison. Some examples of image processing through MATLAB programming language are presented.

4.2 Background

4.2.1 Image Representation

Light microscopy and fluorescence imaging use the visible light with the frequency between 4.28×10^{14} Hz and 7.49×10^{14} Hz or with the wavelength from 400×10^{-9} m to 700×10^{-9} m. They are dedicated to image visualization and analysis of *in vitro* cellular and subcellular structures.

In biological microscopy, to allow the light to pass or to be transmitted through the samples, they are prepared as thin specimens. The main drawback in microscopy is the poor contrast of acquired images, mainly due to the transmitted light through thin samples or reflected light from surfaces showing a high reflectivity. Moreover, the image contrast decreases with tissue depth and the edges into image lose their details. Reducing the number of artifacts is strongly related to the efficient sample illumination that, in turn, is affected by the alignment of all the optical components in the microscope, and by the illumination source as well (Lanni & Keller, 2000; Davidson & Abramowitz, 2002; Evennett & Hammond, 2005; Sluder & Nordberg, 2007; Lorenz et al., 2012; Ashour et al., 2018; Dey et al., 2018; Ripon et al., 2018; Wang et al., 2018;).

Different imaging modes utilized in optical microscopy provided by the objectives' types. Thus, standard bright-field objectives are used as investigation methods, such as bright-field, dark-field, oblique, and Rheinberg illumination techniques; image brightness is determined by the intensity of illumination. When cells in living tissue culture or microorganisms are investigated, phase contrast objectives based on the contrast feature generated into semitransparent samples are the required solution. However, it requires a supplementary device placed at the rear focal plane, which contains both a phase plate with neutral density material and/or optical wave retarders (Nikon, Olympus). There are more types of phase contrast objectives: (1) dark low objectives generate a dark image outline on a light gray background, (2) dark low objectives produce better images in brightfield, (3) apodized dark low phase diminishes the "halo" effects into images obtained in phase contrast microscopy, (4) dark medium gets a dark image outline on a medium gray background, and (5) negative phase contrast or bright medium generates a bright image map on a medium gray background. Differential interference contrast objectives are useful for unstained samples; those regions of the sample where the optical distances increase will appear either brighter or darker, whereas regions where the optical paths decrease will show the inverse contrast (Rosenthal, 2009). Polarized light microscopy produces the best images for samples, showing optically anisotropic character.

Typical images captured in optical microscopy are red-green-blue (RGB) color images. A color image is defined as a digital array of pixel containing color information. In MATLAB, an RGB image is stored as "an m-by-n-by-3 data array that defines red, green, and blue color components for each individual pixel." An RGB array can belong to a double, uint8, or uint16 class. In an RGB array of class double, each constituent spans between 0 and 1. A black pixel is described as (0, 0, 0), and a white pixel as (1, 1, 1). In a RGB space, each pixel is depicted using a third dimension of the data array. In MATLAB space, a pixel (10, 5) in a RGB color image is memorized as RGB (10, 5, 1), RGB (10, 5, 2), and RGB (10, 5, 3), respectively. The MATLAB function is image(RGB). The following functions are used to convert between various image types: im2double (converts an image to double precision), rgb2gray (converts RGB image or color map to gray scale), ind2rgb (converts indexed image to RGB image), or rgb2ind (converts RGB image to indexed image).

On the other hand, in a gray-scale digital image, each pixel is depicted by a scalar value, which is its intensity. The intensity values L depend on the numerical type encoding the image (an 8-bit image has $2^8 = 256$ gray-level intensity values).

The nonuniform background correction or nonuniform illumination correction and primary noise filtering were done by microscope camera settings. After the digital raw images have been acquired and suffered certain preprocessing corrections and rehabilitations provided by microscope software system, the digital content becomes accessible even by visual analysis or by image processing and analysis algorithms.

4.3 Image Processing Using Wavelets

Nowadays, rectangular basis function such as discrete wavelet transform (DWT) with the particular examples, Haar, Daubechies, and Dual-Tree

Complex Wavelet Transform, is commonly used in image-processing field (Gröchenig & Madych, 1994; Porwik & Lisowska, 2004a; Porwik & Lisowska, 2004b; Talukder & Harada, 2007; Ali et al., 2010; Eckley et al., 2010; Moraru et al., 2011; Ali et al., 2014; Bhosale et al., 2014; Sonka et al., 2015). The basic idea behind wavelets consists of separating data into different frequency components (or in a variety of levels) to study each component, to either provide greater understanding of the underlying structure or to store the information carried more compactly. Generally, images contain some redundant data. The DWT is a linear transformation that operates on a vector y of length $= 2^J$, where $J \in N_0$, and transforms it into a numerically different vector of the same length. To do this, for every pair of pixels (a, b), first the following transformations are performed: the average $(b + a)/2$ and the difference $(b - a)/2$. The differences in the transformed vector inform about the fluctuations in the data; larger differences signalize important jumps between pixel values, whereas small differences indicate relatively minor changes in their values. The sum or average indicates the trends in the data. If $c_J = y$ and $i \equiv J$, the wavelet coefficients $d_{i,k}$ and $c_{i,k}$ are constructed by finding the sum and difference between subsequent vectors:

$$d_{i-1,k} = \frac{1}{\sqrt{2}}\left(c_{i,\,2k} - c_{i,2k-1}\right), \quad \text{for } k \in \{1,\, 2,\ldots, 2^{J-1}\} \tag{1}$$

$$c_{i-1,k} = \frac{1}{\sqrt{2}}\left(c_{i,2k} + c_{i,2k-1}\right), \quad \text{for } k \in \{1,\, 2,\ldots,\, 2^{J-1}\} \tag{2}$$

As the most used technique, DWT uses the Haar functions in image coding, edge detection, and analysis and binary logic design (Davis et al., 1999; Munoz et al., 2002; Zeng et al., 2003; Dettori & Semler, 2007; Liu et al., 2009; Yang et al., 2010; AlZubi et al., 2011). For an input constituted as a list of 2^n numbers, the Haar transform separates the signal into two subsignals of half its lengths and pairs up input values, storing the difference and passing the sum. Then, this procedure is repeated in a recursive way, pairing up the sums to provide the next scale. The final outcomes are $2n - 1$ differences and one final sum. A simple example of multiscale analysis consisting of Haar transform for $n = 3$ is provided in Figure 4.1.

Thus, the Haar transform at 1-level is

$$y = (4,6,10,\ 12,8,4,\ 4,\ 6) \overset{H_1}{\to} \left(5\sqrt{2}, 11\sqrt{2},\ 6\sqrt{2},\ 5\sqrt{2}| - \sqrt{2}, -\sqrt{2},\ 2\sqrt{2}, -\sqrt{2}\right)$$

Multiplication by $\sqrt{2}$ ensures the energy conservation.
The Haar transforms at multilevel are

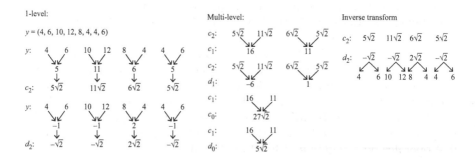

FIGURE 4.1
A numerical example for 1-level, multilevel, and inverse transform of Haar transform for $n = 3$.

$$(c_1|\ d_1\ |\ d_2) = \left(16,\ 11\ |{-}6,\ 1\ |-\sqrt{2}, -\sqrt{2},\ 2\sqrt{2}, -\sqrt{2}\right)$$

$$(c_0|d_0|\ d_1\ |\ d_2) = \left(27\sqrt{2}|5\sqrt{2}|{-}6,\ 1\ |-\sqrt{2}, -\sqrt{2},\ 2\sqrt{2}, -\sqrt{2}\right)$$

The inverse transform is defined as

$$y = \left(\frac{c_1 + d_1}{\sqrt{2}}, \frac{c_1 - d_1}{\sqrt{2}}, \ldots, \frac{c_{N/2} + d_{N/2}}{\sqrt{2}}, \frac{c_{N/2} - d_{N/2}}{\sqrt{2}}\right)$$

The algorithm presented in Figure 4.1 is equivalent with the Haar wavelet. The $d_{i,j}$ are the wavelet coefficients and $c_{0,0}$ determine the output of the transform and allow an alternative way of storing the original sequence.

The DWT provides information in a more discriminating way as it segregates data into distinct frequency components; each frequency component is analyzed using a resolution level adjusted to its scale. The Haar father wavelet is defined as (Vetterli & Kovacevic, 1995; Jahromi et al., 2003)

$$\phi(t) = \begin{cases} 1, & t \in [0,\ 1] \\ 0, & \text{otherwise} \end{cases} \tag{3}$$

The father wavelet coefficients, for a function $f(x)$ and for a required level 2^j, are written as $c_{j,k} = \int f(x) 2^{j/2}\ \phi(2^j x - k) dx$, where $k \in \{1, 2, \ldots, j\}$; the functions denoted by ϕ represent the scaling functions. If $\phi_{j,k}(x) = 2^{j/2}\ \phi(2^j\ x - k)$, then the father wavelet coefficients become:

$$c_{j,k} = \int f(x)\phi_{j,k}(x)dx \tag{4}$$

However, a wavelet transform is defined by two functions. One is the father wavelet which specifies the scale of the wavelet and the second is the mother wavelet which characterizes the shape. The Haar mother wavelet can be defined as follows:

$$\psi(x) = \begin{cases} -1, & x \in [1, \frac{1}{2}) \\ 1, & x \in (\frac{1}{2}, 1] \\ 0, & \text{otherwise} \end{cases} \tag{5}$$

The basic wavelet analysis is based on Haar scaling functions because $\psi(x) = \phi(2x) - \phi(2x - 1)$. The mother wavelet coefficients are computed as

$$d_{j,k} = \int f(x) 2^{\frac{j}{2}} \psi(2^j x - k) dx = \int f(x) \psi_{j,k}(x) dx \tag{6}$$

where ψ denotes the wavelet functions. The Haar father and mother functions are plotted in Figure 4.2.

The decomposition of the original function $f(x)$ can be depicted by a linear combination of mother wavelet:

$$f(x) = \sum_{j,k} d_{j,k} \psi_{j,k} \tag{7}$$

It should be noted that the functions $\phi_{j,k}(x) = 2^{j/2} \phi(2^j x - k)$ and $\psi_{j,k}(x) = 2^{j/2} \psi(2^j x - k)$ form a wavelet family if they are orthonormal base of $L^2(R)$. The constant $2^{j/2}$ is chosen so that the scalar product $\langle \psi_{j,k}, \psi_{j,k} \rangle = 1, \psi_{j,k} \epsilon L^2(R)$.

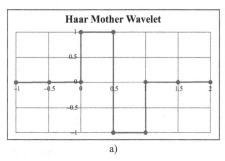
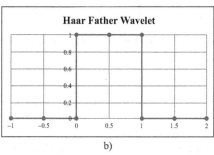

a)　　　　　　　　　　　b)

FIGURE 4.2
The Haar wavelet functions: (a) Haar mother wavelet and (b) Haar father wavelet.

The Haar wavelet has the following characteristics:

- is discontinuous,
- is localized in time; as consequence, it is not perfectly localized in frequency, and
- has low computing requirements.

For these reasons, Haar transforms due to their wavelet-like structures are tools frequently used in image processing. Thus, wavelets, through the mother wavelet coefficients $d_{j,k}$, can be used to analyze the structure (shape) of a function and to obtain information on how much of the corresponding wavelet $\psi_{j,k}$ makes up the original function $f(x)$. If the original function $f(x)$ shows any discontinuity, the mother wavelet $\psi_{j,k}$, which decomposes the function (see eq. 7), will be affected and the values of the coefficients $d_{j,k}$ will clearly indicate this. In addition, the computational complexity is very low, $\mathcal{O}(n)$, and the obtained key information from the function can be summed up concisely. In this case, the efficiency and sparsity of wavelet transform should be mentioned. When image analysis is considered, a natural way to apply the wavelet concept is the 2D wavelets approach (Aldroubi & Unser, 1996). Both computational analysis of the image content and image compression can be addressed.

A digital image $a[m, n]$ is defined in a 2D discrete space. It is derived from an analog image $a(m, n)$ acquired in a 2D continuous space through digitization. This operation divides the continuous image into N ($n = 0, 1, 2, \ldots, N-1$) rows and M ($m = 0, 1, 2, \ldots, M-1$) columns. Another parameter of an image is the number of distinct gray levels (= 2^8 = 256). Starting from 1D mother $\psi_{j,k}$ and father wavelets $\phi_{j,k}$, the 2D discrete mothers are defined using the following three wavelets:

$$\begin{cases} \psi_j^h = \phi_j \otimes \psi_j \ (\textit{horizontal decomposition direction}) \\ \psi_j^v = \psi_j \otimes \phi_j \ (\textit{vertical decomposition direction}) \\ \psi_j^d = \psi_j \otimes \psi_j \ (\textit{diagonal decomposition direction}). \end{cases} \tag{8}$$

The 2D discrete father wavelet is

$$\phi_j^2 = \phi_j \otimes \phi_j \tag{9}$$

where $\otimes$ denotes the outer product: $a \otimes b = ab^T = C$, and $C_{i,j} = a_i \times b_j$.

At this point, in the discrete case, equations (7) and (8) provide a one-to-one representation of the original image in terms of wavelet coefficients. The quantization as well as a simple removal of certain irrelevant coefficients are both useful in image de-noising, image enhancement, and edge detection.

Donoho and Johnstone (1994) and Donoho (1995) developed the mechanism of noise removal in wavelet domain and pointed out that large coefficients match to the signal, and small coefficients contain mostly noise. Weaver et al. (1991) proposed the first approach for noise removal by using an orthogonal decomposition of an image and a soft thresholding procedure in the selection of the coefficients $d_{j,k}$:

$$d_{j,k} = \begin{cases} d_{j,k} - t_i, & d_{j,k} \geq t_i \\ 0, & |d_{j,k}| \leq t_i \\ d_{j,k} + t_i, & d_{j,k} \leq -t_i \end{cases}$$

t_i denotes the threshold and its value depends of the noise at the ith scale, namely, sub-band adaptive. Generally, the coefficients corresponding to noise are characterized by high frequency, and the low-pass filters represent the best solution for noising removal. In thresholding methods, the wavelet coefficients whose magnitudes are smaller than the threshold are set to zero. In the next stage, the de-noised image is reconstructed by using the inverse wavelet transform. The threshold selection is a sensitive process because a higher t_i allows thresholding operation to remove a significant amount of signal energy and to cause over-smoothing. A small t_i means a significant amount of noise that is not suppressed (Chang et al., 2000; Portilla et al., 2003; Moldovanu & Moraru, 2010; Bibicu et al., 2012; Moldovanu et al., 2016). On the other hand, the edge information is presented in the low frequency, so, a balance should be found between these two goals of image processing.

Image enhancement aims to accentuate image features or details that are relevant for a specific task. It is useful where the contrast between different structures in an image is small and a minor change between these structures could be very informative. Wavelet-based enhancement methods use reversible wavelet decomposition and the enhancement is performed by selective modification of wavelet coefficients for highlighting subtle image details (Sakellaropoulos et al., 2003; Zeng et al., 2004; Moraru et al., 2011; Kumar et al., 2013; Patnaik & Zhong, 2014). It is followed by a reconstruction operation.

According to Misiti et al. (2007), the applications of wavelets play an important role in edge detection and image fusion, mainly due to their local character. Edges in images are mathematically described as local singularities that are, in fact, discontinuities in gray-level values (Tang et al., 2000; Misiti et al., 2007; Guan, 2008; Papari & Petkov, 2011; Xu et al., 2012). In fact, for a digital image $a[m, n]$, its edges correspond to singularities of $a[m, n]$ that, in turn, are correlated to the local maxima of the wavelet transform modulus. Discrete Haar wavelet has applications in the location of the specific lines or edges in image by searching all coefficients in the spectral space. This will allow to find all important edge directions in the image. Wavelet filters of large scales are capable of producing noise removal but amplify the uncertainty of the location of edges. Wavelet filters of small scales detect the exact location of edges, but incertitude exists between noise and real edges. Mallat

(1989) demonstrated that large wavelet coefficients characterize the edges but the wavelet representation provide different edge map according to the orientations. These directional limitations of the wavelets can be overcome by edge fusion operation (Pajares & de la Cruz, 2004; Akhtar et al., 2008; Rubio-Guivernau et al., 2012; Xu et al., 2012; Suraj et al., 2014). The result of image fusion is a more desirable image with higher information density that is more suitable for further image-processing operations.

To take advantages from both horizontal and vertical edge map images, the Dempster–Shafer fusion is used (Shafer, 1990; Mena & Malpica, 2003; Seo et al., 2011; Li & Wee, 2014, Moraru et al., 2018). This is a way to fuse information from various level of decomposition by taking into account the inaccuracy and uncertainty at the same time. The final goal is to reduce the uncertainty by combining the evidences from different sources. To apply the Dempster–Shafer theory of evidence, a mass function should be provided. The belief function represents the total belief for a hypothesis and can be integrated into a variational wavelet decomposition framework. Let hypotheses set Ω be composed of j-level of decomposition exclusive subsets Ω_i, then $\Omega = \{\Omega_1, \Omega_2, \ldots, \Omega_n\}$. Each element of this universal set has a power set of $\wp(\Omega)$ and A is an element of this set. The associated elementary mass function $m(A)$ specifies all confidences assigned to this proposition. The mass function $m : \wp(\Omega) \to [0, 1]$ satisfies the conditions:

$$\begin{cases} m(\oslash) = 0 \\ \sum_{A \subseteq \Omega} m(A) = 1 \end{cases}$$

where ϕ is an empty set and $m(A)$ is the belief that certain element of Ω belongs to the set A. The mass function allows defining the interval of upper and lower bounds, which are called belief and plausibility. The belief functions are particularized on the same frame of discernment and are based on independent arguments. The plausibility functions indicate how the available evidence fails to refute A (Scheuermann & Rosenhahn, 2010; Li & Wee, 2014). In the next step, the Dempster–Shafer theory provides a rule of combination that aggregates two masses m_1 and m_2 resulted from different sources into one body of evidence. This rule of combination is a conjunctive operation (AND), and the Dempster's rule or orthogonal product of the theory of evidence is as follows:

$$m_1(A) \oplus m_2(A) = \frac{\sum_{B \cap C = A} m_1(B) m_2(C)}{\sum_{B \cap C \neq \phi} m_1(B) m_2(C)}; \quad (A, B, C \in \Omega)$$

where A, B, and C are event sets generated by the Dempster–Shafer fusion, 1-level decomposition and recomposition, and 2-level decomposition and recomposition, respectively.

The fusion rule decides those detail coefficients that will be suppressed and those that will be fused with the coefficients of another image. The most informative are horizontal and vertical image decomposition structures (de Zeeuw et al., 2012). The fusion technique involves three steps: (1) decompose the images in the same wavelet basis to obtain horizontal, vertical, and diagonal details; (2) recompose horizontal and vertical images; and (3) fuse horizontal and vertical images at j-level. This technique can be used to reconstruct a new image from processed versions of the original image or to construct a new image by combining two different images. Wavelet functions apply derivatives at different levels. In this case, horizontal and vertical details, at various levels of decompositions, are considered as evidences and their outputs are taken as events for Dempster–Shafer fusion.

Below, certain image enhancement and edge detection algorithms are shortly present along with some code listing and examples. The Wavelets and Image Processing Toolbox for MATLAB contains many basic functions used in image-processing applications such as filtering for noise removal, edge detection, and edge enhancement. Haar is a MATLAB library which computes the Haar transform of data. For this chapter purposes, the data are in the form of a 2D array and the transform is applied to the columns and then the rows of a matrix.

The images of cells used in this chapter were downloaded by digital video gallery provided by Olympus website. They were captured using an Olympus fluorescence microscope, for example, fluorescent speckle microscopy, for obtaining high-resolution images of embryonic rat fibroblast cells (A7r5 line) that express a fusion of monomeric Kusabira Orange fluorescent protein to human beta-actin. The test image has a size of 400 × 320 pixels.

4.4 Image-Quality Analysis

Objective image-quality measures are frequently used in image processing, but it is equally important to predict and enhance the quality of the images. There are three approaches for image-quality assessment: subjective, objective, or a hybrid way. The last approach is based on the similarity of the human visual system characteristics coupled with various objective measures. The objective image-quality metrics are fast and reproducible for readers but require expensive software. These metrics can be used in all stages of the processing pipeline, can provide information about the best algorithms, or can be used as an optimization tool in de-noising and image restoration. This chapter focuses on objective image-quality measures such as PSNR and MSE as metrics devoted to the estimation of difference between informational contents when raw or original images are compared with processed images.

The PSNR, given in decibel (dB), is computed as

$$\text{PSNR} = 10 \, \log \frac{N \cdot M \left(\max_{i,j} f(i,j) \right)^2}{\sum_{i=0}^{N-1} \sum_{j=0}^{M-1} \left(g(i,j) - f(i,j) \right)^2}$$

where $f(i, j)$ denotes the original image and $g(i, j)$ is the processed image. The images have $N \cdot M$ size. PSNR compares the restoration results for the same image, and higher PSNR denotes a better restoration method and indicates a higher quality image (Mohan et al., 2013; Moraru et al., 2015). Through this maximization operation in signal-to-noise ratio, the probabilities of missed detection and false alarm in edge detection are minimized. In the wavelets application domain, the main focus was for noise reduction or noise removal in digital images. However, the wavelet transforms show some drawbacks such as oscillations, shift variance, or aliasing (Satheesh & Prasad, 2011; Raj & Venkateswarlu, 2012; Thanki et al., 2018).

The MSE is an error metric that computes the cumulative squared error between the processed and raw/original image (Mohan et al., 2013; Moraru et al., 2015):

$$\text{MSE} = \frac{1}{N \cdot M} \sum_{i=0}^{N-1} \sum_{j=0}^{M-1} \left(g(i, j) - f(i, j) \right)^2$$

Lower portion of MSE indicates lesser errors, but it strongly depends on the image intensity scaling. In other words, the squared difference diminishes the small differences between the two pixels but penalizes the large ones. Wavelet transform as a method for noise reduction or noise removal surpasses this drawback because it has better directional and decomposition capabilities.

To improve the results provided by accepted methods such as PSNR and MSE, which are not accurate to the visual human perception, a hybrid, perceptual method called SSIM index is used. This method measures the similarity between two images (Wang et al., 2004; Moraru et al., 2015). SSIM is a strong evaluation instrument in the case of image fusion because it takes into account luminance and contrast terms (that describe the nonstructural distortions) and the structural distortion that characterizes the loss of linear correlation. For two spatial locations x and y in analyzed images (i.e., two data vectors having nonnegative values expressing the pixel values to be compared), the SSIM is defined as

$$\text{SSIM}\,(x, \, y) = \left(\frac{2 \, \overline{xy} + C_1}{\overline{x}^2 + \overline{y}^2 + C_1} \right)^{\alpha} \left(\frac{2 s_x s_y + C_2}{s_x^2 + s_y^2 + C_2} \right)^{\beta} \left(\frac{s_{x,y} + C_3}{s_x s_y + C_3} \right)^{\gamma}$$

$$= \left(l(x, \, y) \right)^{\alpha} \left(c(x, \, y) \right)^{\beta} \left(s(x, \, y) \right)^{\gamma}$$

where $\bar{x}$ and $\bar{y}$ denote the mean values, s_x and s_y are the standard deviation, and s_{xy} is the covariance. α, β, $\gamma > 0$ are weights on the luminance (l), contrast (c), and structure term (s) of the fidelity measure. C_1, C_2, and C_3 are small positive constants to avoid the division with weak denominator.

4.5 Applications

4.5.1 Decomposition Levels for 2D Haar Wavelet Transform

To analyze the test image, the area of image has been decomposed into 2-level. In the first step, the wavelet decomposition transforms the initial raw image into constituents with single low resolution called "approximation" and "details" as shown in Figures 4.3 to 4.6. The approximation resembles the initial image. In the chosen example, G is approximation and encompasses information about the global properties of the image, A is 1-level decomposition and contains information on the horizontal lines hidden in image, B is 1-level decomposition and comprises information on the vertical lines hidden in image, and C is 1-level decomposition and embraces information about the diagonal details hidden in image (see Figure 4.4). These details components contain mostly noise. The 2-level of decomposition (or higher) is achieved by reiterating the same operations on the approximation. In this case, D is 2-level decomposition and includes horizontal details, E is 2-level decomposition and includes vertical details, and F is 2-level decomposition and embraces diagonal details (as shown in Figure 4.5).

Haar wavelet is the simplest basis and provides adequate localization of the image features. It is also compactly supported, is orthogonal, and is symmetric wavelet.

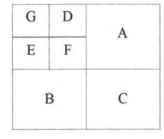
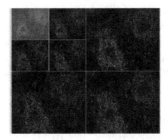

FIGURE 4.3
2-Level wavelet decomposing structure of embryonic rat fibroblast cells. The graph on the left shows the organization of the approximation and details.

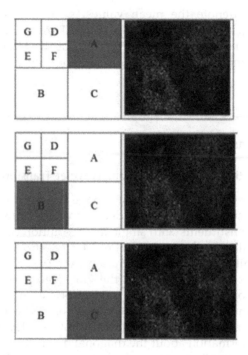

FIGURE 4.4
Approximation and details by directions for 1-level wavelet decomposing structure of test image. The directions are A (horizontal), B (vertical), and C (diagonal).

4.5.2 Noise Removal in Haar Wavelet Domain

The noise removal process consists of the following three steps: (1) wavelet decomposition at 2-level (in this example, but a general *J*-level can be considered) of the raw image; (2) soft (zeroing details and scaling the nonzero coefficients toward zero) or hard (zeroing details) thresholding of the detail coefficients $d_{j,k}$, in the three directions, at a selected level; and (3) reconstruction of the de-noised image. The main problem here is that edges are blurred and an enhancement operation is required.

The de-noising process has been carried out using the Haar wavelet at level 2 and has been performed by Wavelet Toolbox main menu and wavelet 2D. The "code listing 1" depicts the main sequence of the algorithm and the test image to be processed is shown in Figure 4.7. Figure 4.7(a) displays the images of embryonic rat fibroblast cells, whereas Figures 4.7(b) and Figures 4.7(c) show the de-noised image by Haar wavelet and the residuals, respectively. Residuals are the deviation of observations set from the mean of the set (or the variance between computationally obtained coefficients and the experimental results). The following default functions of MATLAB are used: wavedec2 (multilevel 2D wavelet decomposition),

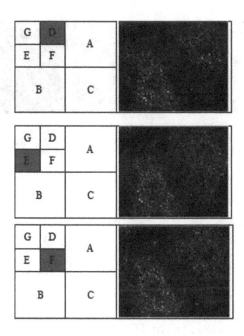

FIGURE 4.5
Approximation and details by directions for 2-level wavelet decomposing structure of test image. The directions are D (horizontal), E (vertical), and F (diagonal).

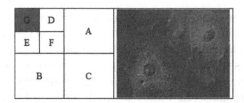

FIGURE 4.6
The approximation component providing information about the global properties of analyzing image.

round (round to nearest integer), isequal (determine array equality), uint8 (convert to 8-bit unsigned integer), and wdencmp (2D de-noising and compression-oriented function).

Code listing 1: Noise removal

```
H = 'haar';
1_2 = 2;
```

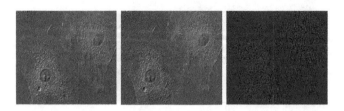

FIGURE 4.7
Noise removal by the Haar wavelet: (a) noisy raw image, (b) de-noised image, and (c) residuals.

```
sorh = 's';
t_S = [5 5 5 5 5 5];
roundFLAG = true;
[c_s,sizes] = wavedec2(X,l_2,H);
[X,cfsDEN,dimCFS]=wdencmp('lvd',c_s,l_2,H,l_2, t_S,sorh);
if roundFLAG
X= round(X);
end
if isequal(class(X),'uint8')
X= uint8(X);
end
```

4.5.3 Image Enhancement by Haar Wavelet

Wavelet transform, as a filter bank, has the ability to get high-frequency information masked in high-frequency sub-images of wavelet transform. The image high-frequency information is used to enhance an image. The de-noising step should be previously done because noise is characterized by high frequency so it exists in high-frequency sub-band images. The de-noised image's sub-bands are obtained by Haar transform. The enhancement is selectively performed by increasing the value of detail coefficients prior to reconstruction through the inverse wavelet transform. In other words, the areas of low contrast will be enhanced more than the areas of high contrast that is equivalent to assign larger gains to smaller detail coefficients.

Some methods for contrast enhancement of digital images are based on the DWT and 1-level decomposition. The reconstructed image based on the inverse DWT IDWT represents the enhanced image. The "code listing 2" contains the default functions of MATLAB software: size (size of object array), dwt2 (single-level discrete 2D wavelet transform), and idwt2 (single-level inverse discrete 2D wavelet transform). The output images are shown in Figure 4.8.

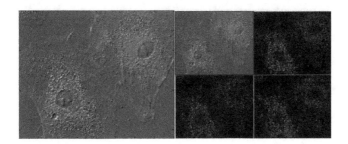

FIGURE 4.8
Image enhancement by Haar wavelet: (a) enhanced image and (b) decomposition at level 1 of de-noised image.

Code listing 2: Image enhancement

```
load rat;
PX = size(X);
[c1,a1,V1,D1] = dwt2(X,'haar');
A = idwt2(c1,a1,V1,D1,'haar',PX);
```

More accurate results are achieved when the analysis is made at the first level. At superior analysis levels, blurring effects appear.

4.5.4 Edge Detection by Haar Wavelet and Image Fusion

Edges are the most important features in digital images. Edges represent the boundaries between the objects embedded into image and background and are detected as discontinuities in gray-level values. Another definition states that an edge is a boundary across which an abrupt change in the brightness of the image exists or a local maximum of the gradient is highlighted.

An edge detector operates on the input image and produces an edge map. Generally, the edge maps provide information about the position and strength of the edges and their orientation. However, the main drawback of edge detectors is smoothing operation, which reduces the contrast of edges and promotes to the poor localization of the weak edges. Wavelet transform allows the edge detection at different scale and for different orientations. Edges are strong localized in an image, but wavelet transforms are suitable to properly detect only to point-singularities. Lines and curve discontinuities are inappropriately extracted. To overcome this issue, the decomposition in horizontal and vertical directions is performed along with the fusion operation between the results of horizontal and vertical directions to achieve more accurate and stable results on edge detection. The Haar wavelet filters are used to find vertical and horizontal edges in an image.

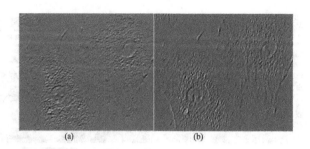

FIGURE 4.9
Edges detection using Haar wavelet: (a) horizontal edge responses and (b) vertical edge responses.

The edge detector presented in "code listing 3" calls the following default functions of MATLAB software: horiH computes the horizontal gradient and vertH transposes the matrix provided by horiH to compute the vertical gradient; integralImage (compute integral image); integralKernel (integral image filter); and imshow (display image). The algorithm provides both 1- and 2-levels of decompositions and recompositions for all possible combinations. The diagonal details were removed.

The 1-level edge images in horizontal and vertical directions are shown in Figure 4.9. Figure 4.10 shows the fused results in the following cases: (a) horizontal edge map details for 1- and 2-levels and their edge fusion, (b) vertical edge map details for 1- and 2-levels and their edge fusion, and (c) horizontal and vertical edge map details at level 1, at level 2, and the combination of both edge fusion for horizontal and vertical direction. The diagonal detail coefficients are neglected because the human visual system has a lower sensitivity to oblique orientations.

Code listing 3: Haar wavelet filters to get vertical and horizontal edges

```
J = imread('rat.jpg');
I=rgb2gray(J);
intImg = integralImage(I);
h_H = integralKernel([1 1 4 3; 1 4 4 3], [-1, 1]);
v_H = h_H.';
h_Reponse = integralFilter(intImg, h_H);
v_Res = integralFilter(intImg, v_H);
imshow(h_Reponse);
imshow(v_Res);
```

4.5.5 Image-Quality Assessment

Successful applications of de-noising via wavelets include the reduction of noise and of other artifacts. SSIMs along with PSNR and MSE have been

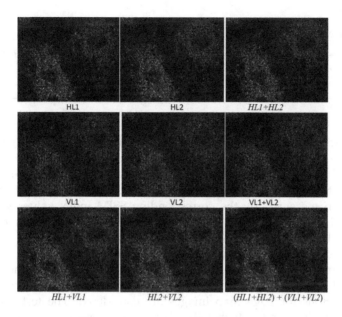

FIGURE 4.10
Fused edge map image provided by Haar wavelet. Both horizontal and vertical edge map details were recomposed following the rules: (a) horizontal details for 1- and 2-levels (HL1 and HL2) and their edge fusion (HL1 + HL2); (b) vertical details for 1- and 2-levels (VL1 and VL2) and their edge fusion (VL1 + VL2); and (c) horizontal and vertical details at level 1 (HL1 + VL1), at level 2 (HL2 + VL2) and for the combination of both edge fusion for horizontal and vertical direction [(HL1 + HL2) + (VL1 + VL2)].

used to evaluate the efficacy of the noise removal. The same enhanced test image was used in this analysis to reveal the effect of a particular choice of wavelet bases in terms of PSNR and MSE values. The enhancement performance is analyzed between the original and enhanced images (with noise removal and enhancement operations) with the same size. Wavelet decompositions and reconstructions of 1- and 2-level were performed on a stack of 20 images. An example of de-noising operation is displayed in Figure 4.7. The average value of PSNR computed between the original and enhanced image with Haar wavelet of 1-level is 36.97 dB. Similarly, when the analysis is performed between the original image and Haar wavelet of 2-level, the average value of PSNR is 36.93 dB. The average values of MSE were computed in the same experimental conditions. For the first case, MSE = 13.25 dB, and for the second case, MSE = 13.24 dB. Comparing the results, no differences are observed between 1- and 2-levels of decomposition for both objective quality metrics; so, the de-noising capability is the same. However, the perceptual quality of the enhanced images is significantly better for enhanced images as noise has been notably reduced, that is, higher PSNR and lower MSE. A partial first conclusion is that the Haar

TABLE 4.1

The average SSIM values of fused images based on three combination rules

	(HL1 + VL1) vs. (HL2 + VL2)	(HL1 + VL1) vs. [(HL1 + HL2) + (VL1 + VL2)]	(HL2 + VL2) vs. [(HL1 + HL2) + (VL1 + VL2)]
SSIM	0.765	0.878	0.844

wavelet transform improves the objective quality of the raw images. However, these very good values of objective metrics do not guarantee avoiding the loss of useful information and preserving the contents of the image. For accuracy, the structural aspects are addressed and the quality of visual information measured by the corresponding SSIM index values is quantified. SSIM captures helpful information on the fidelity with a reference image and estimates the local improvement of image quality over the image space. A value closer to ±1 for the SSIM index value is evidence for a better image quality obtained as a result of image processing in each individual sub-image. Three combinations of the level and directions of decompositions for edge map images are considered. The test conditions are arranged in pairs such that the first is the reference 1- and 2-level of decomposition and the second is a combination between the fused levels of decomposition. An example is provided in Figure 4.10 and the SSIM results are presented in Table 4.1.

From this table, it can be observed that SSIM values have the maximum average value when the image is reconstructed by using the fusion between horizontal and vertical details at level 1. Moreover, the reconstructed images by using the fusion between horizontal and vertical details at level 2 have a good quality. These results conclude that the level of decompositions determines the quality of enhanced images.

4.6 Future Research Directions

In this chapter, only the images provided by the Olympus Microscopy Resource Center galleries are considered. For the future work, it would be useful if the same analysis is performed on images provided by other digital microscope manufacturers. This will enable more efficient image-processing methods specific to the digital pathology area.

To include other factors such as edge sharpness, surface smoothness, and continuity in this analysis or to define an acceptable background noise level, additional models consistent with the human visual system in image-quality prediction could be considered. The perceptual quality prediction model devoted to mean opinion score, mutual information, or picture quality scale can be used.

Furthermore, an analysis on the number of transformation level required in case of pathology applications can be done.

4.7 Conclusions

The Haar wavelet decomposition seems to be one of the most efficient tools to sequentially process and represent an image. The pairs of pixels are averaged to get a new image having different resolutions, but it is required to estimate the loss of information that naturally occurs in this process. The Haar wavelet transform delivers the approximation and detail coefficients at 1- and 2-level of decomposition and it worked like a low- and a high-pass filter simultaneously. The proposed unsupervised machine-learning methods performed better at level 1 of wavelet decomposition and when horizontal and vertical details are fused.

The purpose of this chapter was to analysis the Haar matrix-based method as tool for image processing (such as de-noising, enhancing, edge detection, and edge preserving) and image analysis (such as quality assessment). One of the major advantages of the technique described in this chapter is the multi-resolution approach, which provides a context for a signal decomposition in a succession of approximations of images with lower resolution, complemented by correlative details that sharpen some aspects into the image. To conclude, the main advantages of this approach are as follows: (1) decrease the manual work in the process of analyzing a huge amount of data encompassed in microscopy images; (2) wavelets avoid preprocessing operations such as filtering with dedicated tools, which constitutes a substantial advantage; (3) ensure a good signal-to-noise ratio as well as a good detection rate of the useful information; (4) allow the edge detection at different scale and for different orientations; and (5) improve edge detection by wavelet representation by using the image fusion.

References

Aldroubi, A., & Unser M. (Eds.). (1996). *Wavelets in Medicine and Biology*. Boca Raton, FL, USA: CRC Press.

Ali, S. T., Antoine, J-P., & Gazeau, J-P. (2010). Coherent states and wavelets, a mathematical overview. In *Graduate Textbooks in Contemporary Physics*, New York: Springer.

Ali, S.T., Antoine, J-P., & Gazeau, J-P. (2014). *Coherent States, Wavelets and Their Generalizations*, 2nd edition. New York, USA: Springer.

Akhtar, P., Ali, T. J., & Bhatti, M. I. (2008). Edge detection and linking using wavelet representation and image fusion. *Ubiquitous Computing and Communication Journal*, 3(3), pp. 6–11.

AlZubi, S., Islam, N., & Abbod, M. (2011). Multiresolution analysis using wavelet, ridgelet, and curvelet transforms for medical image segmentation. *International Journal of Biomedical Imaging, 2011,* Article ID 136034, p. 18.

Ashour, A. S., Beagum, S., Dey, N., Ashour, A. S., Pistolla, D. S., Nguyen, G. H., Le, D. N, & Shi, F.. (2018). Light microscopy image de-noising using optimized LPA-ICI filter. *Neural Computing and Applications 29*(12), pp. 1517–1533.

Bhosale, B., Moraru, L., Ahmed, B. S., Riser, D., & Biswas, A. (2014). Multi-resolution analysis of wavelet like soliton solution of KDV equation, *Proceedings of the Romanian Academy, Series A, 15*(1), pp. 18–26.

Bibicu, D., Modovanu, S., & Moraru, L. (2012). De-noising of ultrasound images from cardiac cycle using complex wavelet transform with dual tree. *Journal of Engineering Studies and Research, 18*(1), pp. 24–30.

Chang, S. G., Yu, B., & Vetterli, M. (2000). Spatially adaptive wavelet thresholding with context modeling for image denoising. *IEEE Transactions on Image Processing, 9*(9), pp. 1522–1531.

Davidson, M. W., & Abramowitz, M. (2002). Optical microscopy. *Encyclopedia of Imaging Science and Technology, 2,* pp. 1106–1140.

Davis, G., Strela, V., & Turcajova, R. (1999). Multi-wavelet construction via the lifting scheme. In He, T. X. (Eds.), *Wavelet Analysis and Multiresolution Methods. Lecture Notes in Pure and Applied Mathematics,* New York, USA: Marcel Dekker Inc.

Dettori, L., & Semler, L. (2007). A comparison of wavelet, ridgelet, and curvelet-based texture classification algorithms in computed tomography. *Computers in Biology and Medicine, 37*(4), pp. 486–498.

Dey, N., Rajinikanth, V., Ashour, A. S., & Tavares, J. M. R. S. (2018). Social group optimization supported segmentation and evaluation of skin melanoma images. *Symmetry, 10*(2), p. 51.

de Zeeuw, P. M., Pauwels, E. J. E. M., & Han, J. (2012, February). Multimodality and multiresolution image fusion. Paper presented at the meeting of the 7th International Joint Conference, VISIGRAPP 2012, Rome, Italy.

Donoho, D. L. (1995). Denoising by soft-thresholding. *IEEE Transactions on Information Theory, 41*(3), pp. 613–627.

Donoho, D. L., & Johnstone, I. M. (1994). Ideal spatial adaptation via wavelet shrinkage. *Biometrika, 81*(3), pp. 425–455.

Eckley, I. A., Nason, G. P., & Treloar, R. L. (2010). Locally stationary wavelet fields with application to the modelling and analysis of image texture. *Journal of the Royal Statistical Society: Series C (Applied Statistics), 59*(4), pp. 595–616.

Evennett, P. J., & Hammond, C. (2005). Microscopy overview. In *Encyclopedia of Analytical Science,* (pp. 32–41). Elsevier Ltd.

Gröchenig K., & Madych W. R. (1994). Multiresolution analysis, Haar bases and self–similar tilings of R^n. *IEEE Transactions on Information Theory, 38*(2), pp. 556–568.

Guan, Y. P. (2008). Automatic extraction of lips based on multi-scale wavelet edge detection. *IET Computer Vision, 2*(1), pp.23–33.

Jahromi, O. S., Francis, B. A., & Kwong, R. H. (2003). Algebraic theory of optimal filterbanks. *IEEE Transactions on Signal Processing, 51*(2), pp. 442–457.

Kumar, H., Amutha, N. S., Ramesh Babu, D. R. (2013). Enhancement of mammographic images using morphology and wavelet transform. *International Journal of Computer Technology and Applications, 3,* pp. 192–198.

Lanni, F., & Keller, E. (2000). Microscopy and microscope optical systems. In Yuste, R., Lanni, F., & Konnerth, A. (Eds.), *Imaging Neurons: A Laboratory Manual*, (pp. 1.1–1.72). New York, USA: Cold Spring Harbor Laboratory Press.

Li, X., & Wee, W. G. (2014). Retinal vessel detection and measurement for computer-aided medical diagnosis. *Journal of Digital Imaging, 27*, pp. 120–132.

Liu, X., Zhao, J., & Wang, S. (2009). Nonlinear algorithm of image enhancement based on wavelet transform. In *Proceedings of the International Conference on Information Engineering and Computer Science*, (pp. 1–4). Wuhan, China: IEEE.

Lorenz, K. S., Salama, P., Dunn, K. W., & Delp, E. J. (2012). Digital correction of motion artifacts in microscopy image sequences collected from living animals using rigid and non-rigid registration. *Journal of Microscopy, 245*(2), pp. 148–160.

Mallat, S. (1989). A theory for multiresolution signal decomposition: the wavelet representation. *IEEE Transactions on Pattern Analysis and Machine Intelligence, 11* (7), pp. 674–693.

MATLAB, *Image Processing Toolbox User's Guide*, Retrieved October 30, 2015, from http://www.mathworks.com/help/matlab/creating_plots/image-types.html

Mena, J. B., & Malpica, J. A. (2003). Color image segmentation using the Dempster-Shafer theory of evidence for the fusion of texture. In *The International Archives of the Photogrammetry, Remote Sensing and Spatial Information Sciences* (ISPRS Archives), vol. XXXVI, Part 3/W8, Munich, Germany.

Misiti, M., Misiti, Y., Oppenheim, G., & Poggi, J.-M. (Eds). (2007). *Wavelets and their Applications*. London, UK: Wiley-ISTE.

Mohan, J., Krishnaveni, V., & Yanhui, G. (2013). MRI denoising using nonlocal neutrosophic set approach of Wiener filtering. *Biomedical Signal Processing and Control, 8*, pp. 779–791.

Moldovanu, S., & Moraru, L. (2010). De-noising kidney ultrasound analysis using Haar wavelets. *Journal of Science and Arts, 2*(13), pp. 365–370.

Moldovanu, S., Moraru L., & Biswas A. (2016). Edge-based structural similarity analysis in brain MR images. *Journal of Medical Imaging and Health Informatics, 6*(2), pp. 539–546.

Moraru, L., Moldovanu, S., Culea-Florescu, A. L., Bibicu, D., Ashour, A. S., & Dey N. (2017). Texture analysis of parasitological liver fibrosis images. *Microscopy Research and Technique, 80*(8), pp.862–869.

Moraru, L., Moldovanu, S., & Nicolae, M. C. (2011). De-noising ultrasound images of colon tumors using Daubechies wavelet transform. In *Conference Proceedings of American Institute of Physics*, vol. 1387, (pp. 294–299), Timisoara, Romania: AIP Publishing.

Moraru, L., Moldovanu, S., & Obreja, C. D. (2015). A survey over image quality analysis techniques for brain MR images. *International Journal of Radiology, 2*(1), pp. 29–37

Moraru, L., Obreja, C. D., Dey N., & Ashour, A. S. (2018). Dempster-Shafer fusion for effective retinal vessels' diameter measurement. In Dey, N., Ashour, A. S., Shi, F., & Balas, V. E. (Eds.), *Soft Computing in Medical Image Analysis*. Academic Press Elsevier B&T.

Munoz, A., Ertle, R., & Unser, M. (2002). Continuous wavelet transform with arbitrary scales and O(N) complexity. *Signal Processing, 82*(5), pp. 749–757.

Nikon. *Specialized Microscope Objectives* (n.d.). Retrieved October 30, 2015, from https://www.microscopyu.com/articles/optics/objectivespecial.html

Olympus. *Microscopy resource center* (n.d.). Retrieved October 30, 2015, from http://www.olympusmicro.com/

Pajares, G., & de la Cruz, J. M. (2004). A wavelet-based image fusion tutorial. *Pattern Recogition, 37*(9), pp. 1855–1872

Papari, G., & Petkov, N. (2011). Edge and line oriented contour detection: State of the art. *Image and Vision Computing, 29*(2–3), pp.79–103.

Patnaik, S., & Zhong, B., (Eds.). (2014). *Soft Computing Techniques in Engineering Application. Studies in Computational Intelligence* (vol. 543). Switzerland: Springer International.

Piella, G., & Heijmans, H. (2003). A new quality metric for image fusion. In *Proceedings of International Conference on Image Processing* ICIP 2003, vol. 3, (pp. 173–176), Barcelona, Spain: IEEE

Portilla, J., Strela, V., Wainwright, M. J., & Simoncelli, E. P. (2003). Image denoising using scale mixtures of Gaussians in the wavelet domain. *IEEE Transactions on Image Processing, 12*(11), pp. 1338–1351.

Porwik, P., Lisowska, A. (2004a). The Haar-Wavelet transform in digital image processing: its status and achievements. *Machine Graphics and Vision, 13*, pp. 79–98.

Porwik, P., & Lisowska, A. (2004b). The new graphics description of the Haar wavelet transform. *Lecture Notes in Computer Science* vol. 3039, (pp. 1–8), Springer-Verlag.

Raj, V. N. P., & Venkateswarlu, T. (2012). Denoising of medical images using dual tree complex wavelet transform. *Procedia Technology, 4*, pp. 238–244.

Rajinikanth, V., Satapathy, S. C., Dey, N., & Vijayarajan, R.. (2018). DWT-PCA image fusion technique to improve segmentation accuracy in brain tumor analysis. In *Microelectronics, Electromagnetics and Telecommunications*, pp. 453–462. Singapore: Springer.

Rohit, T., Borra, S., Dey, N., & Ashour, A. S.. (2018). Medical imaging and its objective quality assessment: An introduction. In *Classification in BioApps*, pp. 3–32. Cham: Springer.

Rosenthal, C. K. (2009). Light Microscopy: Contrast by interference. *Nature Milestones | Milestone 8.*

Rubio-Guivernau, J. L., Gurchenkov, V., Luengo-Oroz, M. A., Duloquin, L., Bourgine, P., Santos, A., Peyrieras, N., & Ledesma-Carbayo, M. J. (2012). Wavelet-based image fusion in multi-view three-dimensional microscopy. *Bioinformatics, 28*(2), pp. 238–245.

Sakellaropoulos, P., Costaridou, L., & Panayiotakis, G. (2003). A wavelet-based spatially adaptive method for mammographic contrast enhancement. *Physics in Medicine and Biology, 48*(6), pp. 787–803.

Satheesh, S., & Prasad, K. (2011). Medical image denoising using adaptive threshold based on contourlet transform. *Advanced Computing: An International Journal, 2*(2), pp. 52–58.

Scheuermann, B., & Rosenhahn, B. (2010). Feature quarrels: The Dempster-Shafer evidence theory for image segmentation using a variational framework. In Kimmel, R., Klette, R., & Sugimoto, A. (Eds.), *Computer Vision – ACCV 2010*, (pp. 426–439). Lecture Notes in Computer Science, 6493.

Seo, S. T., Sivakumar, K., & Kwon, S. H. (2011). Dempster-Shafer's evidence theory-based edge detection. *International Journal of Fuzzy Logic and Intelligent Systems, 11*(1), pp. 19–24.

Shafer, G. (1990). Perspectives on the theory and practice of belief functions. *International Journal of Approximate Reasoning, 4*, pp.323–362.

Shamim, R., Chowdhury, L., Ashour, A. S., & Dey, N.. (2018). Machine-learning approach for ribonucleic acid primary and secondary structure prediction from images. In *Soft Computing Based Medical Image Analysis*, pp. 203–221.

Sluder, G., & Nordberg, J. (2007). Microscope basics. *Methods in Cell Biology 81*, pp. 1–10.

Sonka, M., Hlavac, V., & Boyle, R. (2015). *Image Processing, Analysis, and Machine Vision*, CENGAGE Learning, 4th Edition, USA: Global Engineering.

Suraj, A. A., Francis, M., Kavya, T. S., & Nirmal, T. M. (2014). Discrete wavelet transform based image fusion and de-noising in FPGA. *Journal of Electrical Systems and Information Technology, 1*(1), pp.72–81.

Talukder, K. H., & Harada, K. (2007). Haar wavelet based approach for image compression and quality assessment of compressed image. *International Journal of Applied Mathematics, 36*(1), p. 9.

Tang, Y. Y., Yang, L., & Liu, J. (2000). Characterization of Dirac-structure edges with wavelet transform. *IEEE Transactions on Systems, Man, and Cybernetics, Part B: Cybernetics, 30*(1), pp. 93–109.

Vetterli, M., & Kovacevic, J. (1995). *Wavelets and Subband Coding*. Englewood Cliffs, NJ: Prentice Hall.

Wang, Z., Bovik, A. C., Sheikh, H. R., & Simoncelli, E. P. (2004). Image quality assessment: From error visibility to structural similarity. *IEEE Transactions on Image Processing, 13*(4), pp. 600–612.

Weaver, J. B., Xu, Y., Healy Jr., D. M., & Cromwell, L. D. (1991). Filtering noise from images with wavelet transforms. *Magnetic Resonance in Medicine, 21*(2), pp. 288–295.

Xu, P., Miao, Q., Shi, C., Zhang, J., & Li, W. (2012). An edge detection algorithm based on the multi-direction shear transform. *Journal of Visual Communication and Image Representation, 23*(5), pp. 827–833.

Yang, Y., Su, Z., & Sun, L. (2010). Medical image enhancement algorithm based on wavelet transforms. *Electronics Letters, 46*(2), pp. 120–121.

Yu, W., Shi, F., Cao, L., Dey, N., Wu, Q., Ashour, A. S., Sherratt, S., Rajinikanth, V., & Wu, L.. (2018). Morphological segmentation analysis and texture-based support vector machines classification on mice liver fibrosis microscopic images. *Current Bioinformatics*.

Zeng, P., Dong, H., Chi, J., & Xu, X. (2004). An approach for wavelet based image enhancement. In *Proceedings of IEEE International Conference on Robotics and Biomimetics*, Shenyang, China: IEEE.

Zeng, L., Jansen, C. P., Marsch, S., Unser, M., & Hunziker, R. (2003). Four-dimensional wavelet compression of arbitrarily sized echocardiographic data. *IEEE Transactions on Medical Imaging, 21*(9), pp. 1179–1188.

5

Placements Probability Predictor Using Data-Mining Techniques

P. N. Railkar, Pinakin Parkhe, Naman Verma, Sameer Joshi, and Ketaki Pathak
Department of Computer Engineering, Smt. Kashibai Navale College of Engineering, Pune

Shaikh Naser Hussain
Asst Prof. Computer Science Dept., Collage of Science & Arts Al Rass, AL Azeem University, Kingdom of Saudi Arabia

CONTENTS

5.1 Introduction

If huge capital, which is being used for educational data mining, is used for other purposes, a great development can be done. Considering this point, education data mining is applied to predict the performance of students in the proposed system. While doing this, data-mining techniques can be used to identify different perspectives and patterns, which will be helpful to interpret performance of students. It gives a new approach to look and evaluate the students according the performance on a whole new level. Data mining done in education field refers to education data mining. It is used to discover new methods and gain knowledge from database and is useful in decision-making. Here, past and present student records are considered. Different classification algorithms are used to predict the student's placement. Various data-mining tools are used in the process. Knowledge of points, which are helpful in students' placements, are beneficial for both, students as well as management. Management can take more steps, which are beneficial for placement of students. Prediction is a tricky thing. The more the variables are used, the more is the accuracy. Among the classifiers used widely, decision trees are very famous. They have a set of rules that are applied on the dataset and the decision is made. Here, list of tasks to be performed can be given as follows: data preparations, data selection and transformation, implementation of mining model, and prediction model.

The chapter is structured as follows: In Section 2, motivation behind the proposed system is discussed. In Section 3, some widely used terminologies regarding the system are briefly explained. Section 4 reviews the related work of the proposed system. In Section 5, gap analysis in the form of how the proposed system is better than the existing systems' architecture is discussed. In Section 6, system architecture of the proposed system along with the flow of the entire process in the proposed system with the help of sequence diagram is discussed. Section 7 describes the overall scope of the proposed system and the important attributes used in it. In Section 8, algorithm used for the proposed system is explained. Section 9 enlightens us about the minimum software and hardware requirements for the proposed system. In Section 10, other specifications are discussed in terms of advantages, disadvantages, and applications. In Section 11, technologies used in the proposed system are briefly introduced. Section 12 describes the mathematical aspects of the proposed system in terms of mathematical model, set theory analysis, and Venn diagram. In Section 13, improvements that can be done in the proposed system are discussed and the chapter is concluded.

5.2 Motivation

Every year, around 1.5 million engineers graduate from colleges. But only 18.43% of them manage to get jobs. The reason behind their unemployment needs not necessarily be their lack of acquaintance with the concepts in their field. Most of the time, students do not know where they are actually lacking and they get frustrated. The motivation behind the proposed system is to help students understand where they should focus more and what other things they need to follow to get their dream job.

5.3 Terminologies

i. TPO (training and placements officer): Staff under whom all the placement work is done.

ii. GH (GUI handler): An entity to handle the GUI (graphical user interface) considered here for explanation purpose.

iii. PB (profile builder): From the previous data, a particular pattern or profile will be generated by PB with respect to work of algorithm.

iv. DT (dataset training): Algorithm will be trained on the previous data to make prediction.

v. ML (machine learning): Ability of a computer to learn by itself without being explicitly programed.

vi. WEKA (Waikato environment for knowledge analysis): Collaboration of different machine-learning tools.

5.4 Literature Review

In Ref. [1], the authors have discussed a design of a placement predictor, which is specifically tailored using specific attributes. Here, *K*-means classifier is used as a basic classifier. Prediction is made on the basis of classification done on previous information of students. Naïve theorem is used here. Previous information of students was used to get factors like test marks, seminar performance, assignment performance, etc. After analysis, this information was used to get the prediction. In Ref. [2], the authors have used a decision tree algorithm in their prediction model. It was particularly useful for analysis of placement details of the students from the department of computer engineering, Thapar University. It also helped in the placement procedure. In Ref. [3], the authors have proposed a model based on the previously mentioned algorithm [2]. It was more of a comparison of algorithms such as neural networks (NNs), Naïve Bayes, and decision tree algorithms. Various factors considered in comparison were true positive rate, accuracy, and precision. Here, Naïve Bayes was used as the base algorithm. In Ref. [4], analysis and prediction are done using KNN algorithm and fuzzy logic. This analysis is used to construct the model. In Ref. [5], the authors have compared different data-mining tools and classification techniques, and their performance was evaluated. The analysis says that the most appropriate algorithm for predicting students' performance is the multilayer perception algorithm. From the results, it is proven that multilayer perception algorithm is most appropriate for predicting students' performance. This algorithm generally has a higher prediction rate compared to other relative algorithm. Here, different classification algorithms are used to evaluate the performance of students, and it also compares different performance aspects of classification algorithms like MLP, J48, etc. In Ref. [6], the authors have analyzed different techniques used for predicting students' performance in placements.

In the corporate world, data mining plays an important role to both companies as well as educational institutions in their respective fields. It helps in prediction and decision-making regarding the performance of students. Even after graduation, records of passed out students will be maintained, which will be helpful for training the algorithm and improve the accuracy of the results. The proposed system contains the value of attributes in numerical values and predetermined strings, and this stored information can be retrieved when required. However, the system fails to analyze the data due to lack of intelligence.

Naïve Bayes mining is used in the proposed system. It is a web-based application. Extracting the needed data is the purpose behind using this technique. At Amrita Vishwa Vidyapeetham, Mysore, this method was used over 700 students using 19 attributes and the results were winning over other methods such as regression, decision tree, and NN with respect to different attributes and prediction (Table [5.1]). The proposed system considers different factors and provides better accuracy in the prediction results. Students' performances in different attributes are provided as input to the proposed system. In the modern era of digital world, efforts are made to reinvent the education system by not letting it be limited to the traditional lecture system; thereby making the system more useful for the students is the real quest. Most of the times, a huge amount of data are collected but not used to its full extent. As big data is coming into the picture, more adaptive tools are required to handle data with such variety. Mining of data provides state-of-the-art solution in this context by predicting much more accurate results. Currently, it is widely implemented in various areas and serving various purposes like banking, cyber security, etc. Mentioning that, it is still at its early stages in educational data mining. Although research is being done, there are many unexplored areas. Moreover, these researches lack a unified approach. This chapter presents the evolution of data mining (2002–2014). In educational data mining, there is hybrid of different techniques such as statistics, machine learning, and data mining which are used to get the useful extract from the data dump. The proposed system helps the students to choose more suitable career path for them on the basis of the current performance and the attributes considered in the system to make the prediction. Clustering and classification are used as the backbone in the prediction process. Decision tree and Naïve Bayes are the classification algorithms and K-means is the clustering algorithm. When applied on the same data, it was found that K-means had higher accuracy. The work done is helpful for students to know their placement chances beforehand and work accordingly later. Comparison between transductive and inductive learning is explored via semi-supervised learning. Generalization of rules is done in transductive learning. It is not necessary in inductive learning [7]. Classification is done by many methods or combination of many methods. For example, classification can be done using the fuzzy systems. It can be done by artificial neural networks (ANNs) or by a hybrid of both mentioned methods. According to research, ANN has more accuracy than the conventional methods used for classification. ANN has a very high precision rate and is very flexible in nature with the datasets used. Apart from this, fuzzy logic is preferable when data are more imprecise. For fuzzy, rules need to be set in a very detailed manner, and defuzzification process has a great deal of importance here. When ANN is considered, it does not examine the recognition of patterns. In this aspect, fuzzy is more reliable as it can show how the decision has been made. The drawback with fuzzy model is that it is less flexible to new data to be processed at later stages. The hybrid of both, ANN and fuzzy, can be used to realize the benefits in both methods. There has been a research conducted to use the neuro-fuzzy model in medical science. Classification techniques like this are essential for better performance [8]. In the last 10 years, many classifier systems have been researched deeply for increasing the precision rate. The main issue arises when the suitable pivot function has to be selected. It has to be mapped with the data pool, subsets, and classifiers. This selection is dependent currently on a selection function which considers both accuracies and the results and the diversity of the dataset. Now, the association of classifiers is calculated using minimum redundancy maximum relevance, which takes the points mentioned before into consideration [9]. NN is a machine-learning model which performs similar to that of the human brain, as the name suggests. It has three layers, namely, input, hidden, and output layers. The ANN has one more layer called raining layer. It means that it can perform and create random functions which are not linear with input and covert them for the required pattern [10]. Medical science is adapting machine learning at an evergrowing rate because of the great accuracy that different classification methods are showing. With the initial dataset collection which is convenient to get from medical records of patients with their permission, specified rules can be set for the prediction [11]. The results obtained using Euler method is compared with the results obtained from the data which excluded the Euler number feature. The performance evaluation is done by comparing features with and without the Euler number. The effectiveness and accuracy of the features were computed by calculating the sensitivity, specificity, and the positive predictive value [12]. Using NN provides speed in segmentation, counting, and classification. Various analysis and consistency employed different NN types to determine the viability of direct classification. Moreover, for classification, the NN classifier is recommended [13].

The main trigger is using region of interest. Initially, it was done manually. So irrespective of the working of the model, possibility of introducing bias was quite natural. This can be overcome using Haralick features in backpropagation of NNs where an optimization function is used to minimize maximum errors. This results in high accuracy with great effective details [14]. The vectors are used as a driving function to increase the accuracy of the results. It is most commonly done by an optimization function which performs the major functionality of minimizing errors. The main disadvantage of using traditional learning algorithms is that it gives the results prematurely. Therefore, accuracy is an issue here. As ANN uses backpropagation, local optimizations are performed at the learning stage itself, which is beneficial for the accuracy [15]. NNs are becoming the new and advanced way of machine learning. The most adaptive and famous modeling tool is thought of as the ANN. It can work on less amount of data and still provide the accurate results. It is considered as a biological neuron because the working of both is kind of similar. The main advantage of using ANN is that it is adaptive in its learning stage itself and can handle similar functional relationships [16].

5.5 Gap Analysis

Gap analysis is described in Table 5.1:

TABLE 5.1

Related Work versus Proposed System

Reference	Gap analysis
[1]	By using C4.5 decision tree algorithm, error rates will be reduced due to inherent nature of the algorithm
[2]	C4.5 decision tree algorithms are advance version of decision tree algorithms, that is, ID3 algorithm
[3]	C4.5 decision tree algorithm gives more accuracy
[4]	C4.5 decision tree algorithm gives more precision rate
[5]	C4.5 decision tree algorithm consists of both more accuracy and precision rate. As a result, it works more efficiently

K-means is a very basic classifier and its error rates are high [1]. This is tackled using C4.5 algorithm. The number of attributes in consideration also matters. They cannot be too high or too low. If they are less, there is a scope for improvement in accuracy of results [2]. Naïve Bayes algorithm has its advantages, but it has less accuracy [3]. It also has low precision rate [4] than C4.5 algorithm. NNs have more accuracy, but the precision rate is low [5]. C4.5 algorithm tends to tackle all these issues.

5.6 System Architecture and Overview

The architecture for the designed system is described below:

It is divided for three different types of users. First user is a student. He/She can use the interface to find out how far he/she can go in the placements rounds, individual performance, and scope for improvement as well. Second user is the TPO. He/She can update the company information. Along with that, access to students' individual progress and overall department performance in the placements can be viewed. Basically, the role of a TPO is to act as a link between company and students. Third user is admin. The role of admin is handling the overall functionality of the system. Web-based GUI will help a user have a friendly interface. Data parser is for parsing the new input. Data selection and transformation will work on past input. The classifier used here is the C4.5 decision tree algorithm. The prediction will be made and entered into the database system. The results will be known to individuals on the GUI again. A sequence diagram depicts the communication or interactions between different objects which belong to the same entity.

In simple words, the system can be explained as follows.

A web-based system which will be used for placement prediction is designed here. There are three people who are included in this system, namely, student, admin, and the TPO. First TPO can login to the server with valid credentials. After login, the TPO will be able to add placement records and company details as depicted in Figure 5.1. The server will fetch the placement records and company details. Server can validate the placement record and company

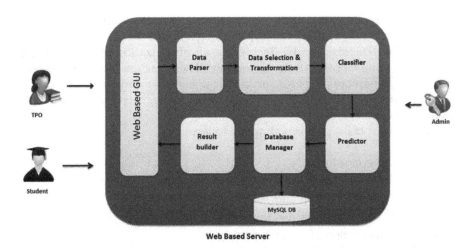

FIGURE 5.1

Architecture of proposed system (brief system architecture).

details. Server will update the database. Student will complete the registration process from the server. Student can update his profile. Server will fetch the student profile and update into student database. Student will be able to view the company details such as company name, company location, criteria, and package. Student can check placement criteria. Server will fetch previous placement record and student's current record. After fetching the records, server will apply the algorithm for classification and prediction. The decision on whether the student can be placed in these companies or not will be determined accordingly. If the prediction is true, then the server will send the notification to students. If the prediction is false, then the server will inform the students that the criterion of company does not match your profile. The sequence diagram is useful to get the interaction between entities considering their order of execution. The flow of the system is given in Figure 5.2.

In Figure 5.3, the work of the three main entities in the proposed system is discussed individually. Basically, the admin is responsible for adding all the entities to the system, and by doing so, access is granted to individual entities. Students can view their statistics company-wise. TPO has access to all placement details. TPO can also view these details graphically for better understanding of the entire scenario. If at all any update is required in the system and also in the information regarding staff or students, it has to be done via admin access which gives admin the most important role in this system. Moreover, students will be notified via SMS system if any new company is available for placement purpose.

5.7 System Details

The success percentage can be improved progressively by knowing certain factors regarding the undergoing subject which here is placements. These factors can be decided at an organization level which can vary from one organization to another. These factors may include the personal behavioral aspects, academical aspects, and general aspects, which are needed to grow successfully in the corporate world (for instance, communication).

Here, major tasks performed are as follows:

   i. Data preparation:
      To decide which dataset is to be used.

  ii. Data transformation and selection:
      Here, fields required for data mining are selected.

 iii. Implementation of mining model:
      Here, C4.5 classification algorithm is applied on the final dataset.

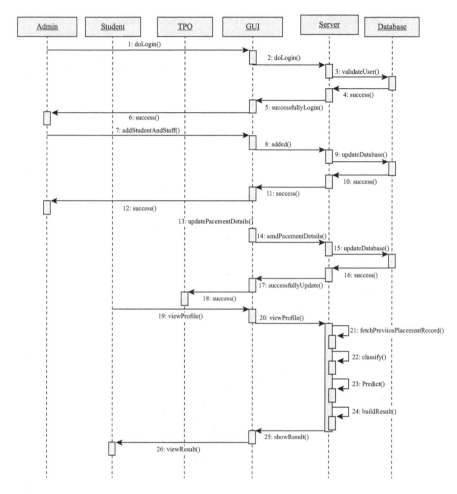

FIGURE 5.2
Sequence diagram for proposed system (flow of activities in the proposed system).

iv. Predictive model:
This model will give the prediction to whether students will get placed in a particular company or not.

Here, the proposed system mainly differs from others in the efficiency of the algorithm and the scope of attributes considered. In addition, some of the corner cases are also considered such as the ability of a student to handle pressure which can be judged by his/her nonacademic performance as well. The order of attributes in which the decision tree is made is not important, but

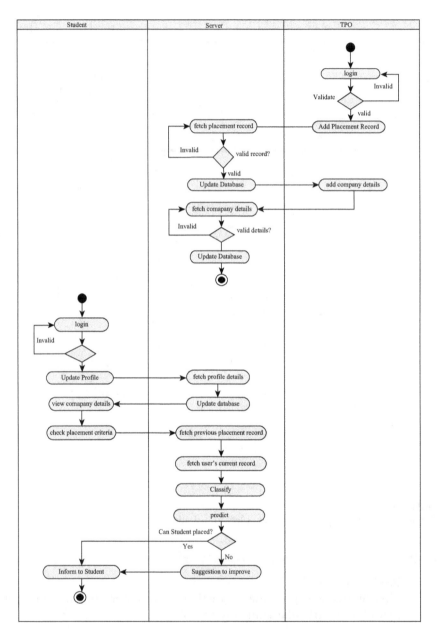

FIGURE 5.3
Activity diagram for proposed system (dynamic control structure for the proposed system).

consideration of all attributes is of prime importance. If by considering all attributes, a tree reaches its leaf node for a particular student for a particular company, his/her probability to get placed in that company automatically increases. However, giving 100% probability is a hurdle because there are runtime factors to be considered as well. The proposed system helps students to get a view of their potential according to the requirement and expectation of various companies and helps the management analyze the whole process. All the attributes here have either numerical values or predetermined string values. They are easy to process. Selection of attributes is a major issue. The algorithm may slow down with increased amount of data and increased number of attributes as well. Hence, choice of appropriate attributes which are actually helpful in the placement process is crucial. Furthermore, some additional skills which are useful in application-oriented nature of jobs are needed to be considered. Here, they are considered as follows:

i. **BE result** – This attribute consists of marks obtained in degree. It includes aggregate marks of students in all the years. It is divided into three classes, such as first ≥65%, second ≥55%, and third <55%.

ii. **Marks of students obtained in secondary and higher secondary examinations.**

iii. **Marks of students obtained in placement-related exam, which includes following fields:**
  1. Quantitative aptitude.
  2. Logical aptitude.
  3. English proficiency.
  4. Particular branch-related technical aptitude fields.

iv. **Semester** – Semester performance of a student is evaluated. In each semester, seminars are conducted for evaluating the performance of students. This performance can be categorized as per need. Here, it is classified as given below:
  1. **Needs improvement** – Communication skills (CSs) and management skills are low.
  2. **Good** – Both presentation and management skills are fine.
  3. **Average** – Management and CSs are acceptable but can be improved.

v. **Lab work** – Technical expertise can be judged to a certain extent. It can be categorized as per need. Here, it is classified as given below:
  1. **Yes** – Candidate has completed the work regarding practical assignments in laboratory efficiently.
  2. **No** – Candidate has not completed the work regarding practical assignments in laboratory efficiently.

vi. **Graduation background** – It defines the student's background in education.

vii. **CS** – It defines the talking skills of a student and is divided into three classes:

   1. **Needs improvement** – CSs are low.

   2. **Good** – CSs are fine.

   3. **Average** – CSs are acceptable but can be improved.

viii. **Placement** – It defines if the student has been placed before or not. It has two values:

   1. **Yes** – Student is placed in a company.

   2. **No** – Student is a fresher.

ix. **Backlog** – This field is used to check whether the student has any active or dead backlogs. Moreover, it includes the number of backlogs.

x. Event management.

xi. Extraordinary performance in various sports.

xii. Lifetime achievements (being a topper).

xiii. Number of internships done which is relevant to the specified job.

xiv. Represented the institution in different kinds of extracurricular activities like singing, dancing, stage play, etc.

5.8 Algorithm Used

The proposed system uses C4.5 decision tree algorithm [17]. Pseudocode of the algorithm is shown below:

Input: an attribute valued dataset D

1. Tree = { }
2. If D is "pure" OR other stopping criteria met then
3. Terminate
4. end if
5. for all attribute a_D do
6. Compute information-theoretic criteria if we split on a
7. end for
8. a_{best} = Best attribute according to above computed criteria
9. Tree = Create a decision node that tests a_{best} in the root
10. D_v = Induced subdatasets from D based on a_{best}

11. For all D_v do
12. Tree = C4.5 (D_v)
13. Attach Tree$_v$ to the corresponding branch of Tree
14. end for
15. return Tree

Above pseudocode is explained briefly as follows:

Step 1: Attribute values are considered and decision tree is constructed.

Step 2: Items which have similar properties are segmented together.

Step 3: There is a separate method defined for selection of attribute which gives the best node to expand and explore.

Step 4: The abovementioned method also considers information gain.

Step 5: Classification can be calculated using following formula:

$$\text{Info}(D) = \sum_{j=1}^{v} \frac{|D_j|}{|D|} \times \text{Info}(D_j)$$

The term $|D_j|$ is used to describe weight of the partition j.
Info(D) is the actual information gain that is computed by calculation and can be used further to explore the tree.

Step 6: Gain is useful to construct decision tree. It helps in branching.
Gain(A) = Infor(D) − Info(D_j)
Hence, gain ratio is taken into consideration for processing.

Step 7: Splitting information can be done as:

$$\text{SplitInfo}(D) = -\sum_{j=1}^{v} \frac{|D_j|}{|D|} \times \frac{\log(|D_j|/|D|)}{\log 2}$$

Step 8: The maximum information gain having node is used for expansion.

Step 9: The procedure is repeated to construct the decision tree with all considered attributes.

5.9 Software and Hardware Specifications

Minimum software and hardware specifications are listed below:

Minimum processor requirement is Intel Core2Duo, Pentium 4/i3. Along with that, minimum speed of the processor should be 2.4 GHz. RAM should be at least 1 GB. Minimum capacity of hard disk should be 50 GB. Operating system compatible is Windows 7 and above. Java 8 is used to design front end. In back end, MySQL 6 is used. Development tool used is JDK 1.8. Integrated Development Environment used to develop the system is Eclipse Luna. So, any Eclipse version above Luna is also compatible. Serer used here is Tomcat 8.

5.10 Other Specifications

These are discussed in the form of advantages, disadvantages, and applications.

i. **Advantages:** It is beneficial for the students to see their performance via using this system. Students can know where they are lagging behind. Students can improve their performance accordingly; thereby, the chance of placement also increases. Students can prioritize their improvement in various fields according to the company's requirements. At the management level, placement records of students can be viewed in a simple format.

ii. **Disadvantages:** Accuracy increases slowly. There is dependency in data between the prediction and past records. The system may get improved if better performing algorithms are used. The system may never get a 100% accurate result due to many factors such as improvisation in the selection process of a company, on time performance of students, etc. If more number of attributes are considered, it will be very time consuming to process the generated decision tree.

iii. **Applications:** Students can use this application for improving their chances in the placement process. Students can know where they stand at the moment in the placement process. The prediction may help students to perform better. At managerial level, all records can be seen in a simple manner. All students' records can be seen at a time via graph which may help the institute to improve their placement records by conducting various activities for students in the sectors in which they are lagging.

iv. **Technologies used:**

Following are the technologies used:

1. **MySQL:** It is a very simple and open-source database tool. It falls under the category of RDBMS. Here, we have used it to store student and company information.

2. **Java**: Java is an evolutionary programming language which can be used for various purposes. It is very useful in network-based client server applications. It has a feature called WORA which is an acronym for "Write Once Run Anywhere." That means it is feasible to use on different platforms. Here, its advanced features have been used to implement the UI and establish connectivity with the back end.

3. **WEKA**: WEKA is basically the culmination of different machine-learning algorithms and features. It is very easy to understand and use as well. It has some predefined methods which can be used for classification and prediction [18].

5.11 Mathematical Model

Set theory analysis is given below:

Let S be the "Placement Prediction."

S = { }

Let S be categorized into different subsets as follows:

S = {S1, S2, S3, S4, S5, S6}

S1 = Graphical user interface handling system (GH)

S2 = Dataset training (DT)

S3 = Profile builder (PB)

S4 = Graphical output (GO)

S5 = Machine learning (ML)

S6 = Company details (CD)

First task is to identify the inputs.

Inputs = {X1, X2, X3, ..., Xn}
Here,

X1 = Profiles (Student, TPO, admin, etc.)

X2 = Company dataset

Let the output set be identified as O.

Outputs = {Y1, Y2, Y3, ..., Yn}

Here,

Y1= Placement prediction

Set theory is described below:

Serial number	Description	UML description
1	**Problem description**	
	Let S be module which contains following submodules, S1 = GUI handler (GH) S2 = Dataset training (DT) S3 = Profile builder (PB) S4 = Graphical output (GO) S5 = Machine learning (ML) S6 = Company details (CD)	S holds the list of modules in the proposed system
2	**Activity**	
	2.1 Activity I **Login Process** Let S1 be a set of parameters which can be used for login S1 = {u/n, pwd} where u/n: Username pwd: Password	If the given input username/ password are correct, allow user to navigate to home page Else, show proper error message

Condition/Parameter	Operation/Function
If login == valid login	f1: Proceed ()
Else.	Throw error

	2.2 Activity II **Upload data process** Let S2 be a set of dataset upload: S2 = {student_id, marks, hobbies, achievements, apti-marks} where student_id: Student ID marks: Marks of different subjects hobbies: Hobbies achievements: Achievements in different area apti-marks: Aptitude marks	Students will upload their profile data which will be helpful to predict the placement chances of that student in the particular company

Condition/Parameters	Operation/ Function
Dataset	f1: doTraning ();

If (dataset is valid)
 Do Training
 Else
 Throw error
	f2: validate-Dataset ();
	 f3: error ();

	2.3 Activity III **Show output** Let S3 be the set which is useful for graphical user interface	The system will show placement prediction in graphical format and system will send SMS to student and TPO as well

(Continued)

(Cont.)

Serial number	Description	UML description
	S3: {prediction, notification} where Prediction: Placement prediction value notification: SMS notification	
3.	**Venn diagram** As described above, entire process can be described as: Input Output (Student Profile) (Placement Prediction) 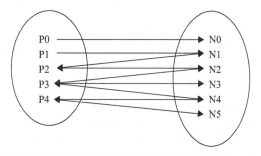	

5.12 Conclusion

The main objective of the proposed system is to develop a system which will give precise and to-the-point prediction. Using the algorithm of C4.5, the students are classified based on their performance attributes. Information gain of each and every attribute is calculated and a decision tree is developed. Here, information gain is an important parameter. The attribute having the maximum information gain will be the root node and will be expanded and explored further. Then, for sub-nodes, same procedure is followed. This process continues till a leaf node is obtained. So, the decision tree made may differ from one individual to another, but it will make sure that all the attributes are considered and the end result format will be the same for each end user regardless of their results. Thus, this domain-specific predictor will give accurate results with more number of records and attributes as well.

Future work consists of applying the proposed system for actual use. It can be used in different applications just by making few changes in parameters. The reports from all applications can be analyzed and can be further used for enhancements. The classification algorithms can be improved in terms of performance and scalability which will lead to the overall performance enhancement of system.

References

[1] Devasia, Tismy, T. P. Vinushree, and Vinayak Hegde. "Prediction of students' performance using Educational Data Mining", Dept. of Computer Science, Amrita Vishwa Vidyapeetham University, Mysuru campus, IEEE 2016.

[2] Pruthi,Karan, and Dr. Prateek Bhatia. "Application of Data Mining in predicting placement of students", Dept. of Computer Science, Thapar University, IEEE 2015.

[3] M. V., Ashok, and Apoorva A. "Data Mining Approach for Predicting Student and Institution's Placement Percentage", Dept. of Computer Science, Bangalore, IEEE 2016.

[4] Sheetal B, Mangasuli, and Prof. Savita Bakare. "Prediction of Campus Placement Using Data Mining Algorithm Fuzzy logic and K nearest neighbour", Department of Computer Science and Engineering, KLE DR M S Sheshgiri College of Engineering and Tech, Belgaum, India, IJARCCE 2016.

[5] Ramesh,V., P. Parkavi, and P. Yasodha. "Performance Analysis of Data Mining Techniques for Placement Chance Prediction", IJSER 2011.

[6] Dwivedi, Tripti, and Diwakar Singh. "Analysing Educational Data through EDM Process: A Survey", Department of Computer Engineering, BUIT, Bhopal, IJCA 2016.

[7] Zemmal, Nawel, Nabiha Azizi, Nilanjan Dey, and Mokhtar Sellami. "Adaptive semi supervised support vector machine semi supervised learning with features cooperation for breast cancer classification." *Journal of Medical Imaging and Health Informatics* 6, no. 1 (2016): 53–62.

[8] Ahmed, Sk Saddam, Nilanjan Dey, Amira S. Ashour, Dimitra Sifaki-Pistolla, Dana Bălas-Timar, Valentina E. Balas, and João Manuel RS Tavares. "Effect of fuzzy partitioning in Crohn's disease classification: a neuro-fuzzy-based approach." *Medical & Biological Engineering & Computing* 55, no. 1 (2017): 101–115.

[9] Cheriguene, Soraya, Nabiha Azizi, Nilanjan Dey, Amira S. Ashour, Corina A. Mnerie, Teodora Olariu, and Fuqian Shi. "Classifier Ensemble Selection Based on mRMR Algorithm and Diversity Measures: An Application of Medical Data Classification." In *International Workshop Soft Computing Applications*, pp. 375–384. Springer, Cham, 2016.

[10] Dey, Nilanjan, Amira S. Ashour, Sayan Chakraborty, Sourav Samanta, Dimitra Sifaki-Pistolla, Ahmed S. Ashour, Dac-Nhuong Le, and Gia Nhu Nguyen. "Healthy and unhealthy rat hippocampus cells classification: a neural based automated system for Alzheimer disease classification." *Journal of Advanced Microscopy Research* 11, no. 1 (2016): 1–10.

[11] Bhattacherjee, Aindrila, Sourav Roy, Sneha Paul, Payel Roy, Noreen Kausar, and Nilanjan Dey. "Classification approach for breast cancer detection using back propagation neural network: a study." In *Biomedical image analysis and mining techniques for improved health outcomes*, p. 210, 2015.

[12] Maji, Prasenjit, Souvik Chatterjee, Sayan Chakraborty, Noreen Kausar, Sourav Samanta, and Nilanjan Dey. "Effect of Euler number as a feature in

gender recognition system from offline handwritten signature using neural networks." In *2015 2nd International Conference on Computing for Sustainable Global Development (INDIACom)*, pp. 1869–1873. IEEE, 2015.

[13] Kotyk, Taras, Amira S. Ashour, Sayan Chakraborty, Nilanjan Dey, and Valentina E. Balas. "Apoptosis analysis in classification paradigm: a neural network based approach." In *Healthy World Conference*, pp. 17–22. 2015.

[14] Samanta, Sourav, Sk Saddam Ahmed, Mohammed Abdel-Megeed M. Salem, Siddhartha Sankar Nath, Nilanjan Dey, and SheliSinha Chowdhury. "Haralick features based automated glaucoma classification using back propagation neural network." In *Proceedings of the 3rd International Conference on Frontiers of Intelligent Computing: Theory and Applications (FICTA) 2014*, pp. 351–358. Springer, Cham, 2015.

[15] Chatterjee, Sankhadeep, Sarbartha Sarkar, Sirshendu Hore, Nilanjan Dey, Amira S. Ashour, Fuqian Shi, and Dac-Nhuong Le. "Structural failure classification for reinforced concrete buildings using trained neural network based multi-objective genetic algorithm." *Structural Engineering and Mechanics* 63, no. 4 (2017): 429–438.

[16] Chatterjee, Sankhadeep, Sarbartha Sarkar, Sirshendu Hore, Nilanjan Dey, Amira S. Ashour, and Valentina E. Balas. "Particle swarm optimization trained neural network for structural failure prediction of multistoried RC buildings." *Neural Computing and Applications* 28, no. 8 (2017): 2005–2016.

[17] https://www.researchgate.net/publication/267945462_C45_algorithm_ and_Multivariate_Decision_Trees

[18] https://www.cs.waikato.ac.nz/ml/weka/

[19] MySQL – The world's most popular open source database, http://www.mysql.com/

[20] http://enos.itcollege.ee/~jpoial/allalaadimised/reading/Advanced-java.pdf

6

Big Data Summarization Using Modified Fuzzy Clustering Algorithm, Semantic Feature, and Data Compression Approach

Shilpa G. Kolte
Research Scholar, University of Mumbai

Jagdish W. Bakal
Professor, S S. Jondhale College of Engineering, Dombavali, India

CONTENTS

6.1 Introduction

Due to huge information produced by numerous sources, interpersonal organizations, and cell phones, we get huge volumes of data known as Big

Data. The gigantic development in the size of data has been seen recently as a key factor of the Big Data. Big Data can be characterized as high volume, velocity, and variety. Computational infrastructure for analyzing Big Data is large. Big Data is the term for a gathering of informational indexes so expensive and complex that it becomes hard to process utilizing conventional information preparing applications. There are three Vs: volume (a lot of information), variety (incorporates distinctive sorts of information), and velocity (always collecting new information) [1], which portray Big Data. Data turn out to be big when their volume, variety, or assortment surpasses its capacities of frameworks to store, analyze, and process them. As of late, general understanding is more prominent after adding two more Vs. Hence, Big Data can be explained by five Vs, namely, volume, velocity, variety, veracity, and value [2]. Big Data is not just about heaps of information, they are really another idea of giving a chance to locate knowledge into the current information. There are numerous usages of Big Data, such as business, innovation, media transmission, drug, human services, and administrations, bioinformatics (hereditary qualities), science, online business, the Internet (data seek, interpersonal organizations), and so forth. Big Data can be gathered from computers, as well as from billions of mobile phones, web-based social networking sites, distinctive sensors introduced in autos, transportation, and numerous different sources. Large data are simply being created at speeds greater than that of which they can be prepared and analyzed. The new difficulty in data mining is that substantial volumes and distinctive assortments must be considered. The basic techniques and devices for data preparing and examination cannot oversee such measures of data, regardless of whether capable Personal Computer (PC) groups are used [3, 4]. To study Big Data, numerous data mining and machine-learning techniques as well as advancements have been found and improvised. Thus, Big Data yields new data and capacity components, which in turn require new strategies for investigation. While managing Big Data, data clustering becomes a major issue. Clustering methods have been connected to numerous critical issues [5], for instance, to find social insurance drifts in understanding documents, to find out duplicate entries in address records, to get new classes of stars in cosmic information, to partition information into clusters that are significant and valuable, and to group a large number of documents or website pages. To provide solutions to these applications, many clustering algorithms have been created. Recent clustering algorithms have a couple of impediments. Most calculations require checking the enlightening record for a couple of times; therefore, they are not suitable for Big Data grouping. There are a lot of usages, in which, to an extraordinary degree, colossal or immense educational accumulations ought to be examined, which in any way are excessively costly, making it impossible to ever be set up by regular grouping strategies. Document summarization acts as an instrument to gain speed, by accepting the accumulation of text documents, and has various genuine applications. To generate powerful summary of expansive text collection, semantic similarity and clustering can be used proficiently [7]. Summarizing expansive volumes of content is a difficult and wearisome job. Process of text summarization includes escalated content preparing and calculations to produce the summary. MapReduce is a demonstrated condition of workmanship innovation for taking care of Big Data [8]. Content summarization is one of the goals and testing issues in content mining. It gives different points of interest to customers and different beneficial bona fide applications that can be made using content mining. In content outline, far-reaching collections of substance documents are changed to a diminished and limited substance record, which addresses the procedure of the principal content aggregations. The calculation which plays out the endeavor of content rundown is known as content summarizer. The content summarizers are grouped into two categories, which are single-report summarizer and multi-report summarizers. In single-report summarizers, a lone tremendous substance chronicle is shortened to another single record rundown; however in multi-report synopsis, a course of action of substance documents (multi-records) are compacted to a singular file plot. Multi-report diagram is a procedure used to gather various substance chronicles and is used for seeing tremendous substance record gatherings. Multi-report rundown delivers a preservationist outline by expelling the huge sentences from a group of records in view of the file subjects. In recent years, researchers have given much thought toward making record rundown techniques. Many rundown systems are proposed to make outlines by removing the basic sentences from the given accumulation of records.

Multi-record rundown is used for cognizance and examination of tremendous file gatherings; the significant wellsprings of these accumulations are news documents, sites, tweets, website pages, ask about papers, web question things, and specific reports open over the web and distinctive spots. A couple of instances of the usages of the multi-record synopsis are researching the web questions for helping customers in examining [9] and making outline for news articles [10]. Report handling and rundown age in an enormous substance file gathering is computationally awesome endeavor. In the time of Big Data, there is a need of new calculations for shortening the tremendous substance collections rapidly, where size of data aggregations is high. In this chapter, we proposed the MapReduce structure-based framework to make the synopsis from sweeping substance collections. For the experiment, UCI machine-learning informational indexes reveal that the computational time for outlining huge content gathering is fundamentally lessened using the MapReduce structure, and MapReduce offers adaptability to obliging tremendous substance collections for shortening. Single-record rundown is anything but difficult to deal with, since only a solitary substance report ought to be dismembered for outline; however, dealing with multi-record synopsis is an astounding and troublesome issue where different gigantic substance documents ought to be explored to make a decreased and illuminating synopsis from it. As the amount of archives augments in multi-record outline, the summarizer gets more inconveniences in creating rundown. If the summarizer contains more gainful and correlated negligible depiction of gigantic substance aggregations, it is said to be incredible, considering semantic equivalent terms give benefits with respect to delivering more relevant synopsis. In this chapter, the issues with multi-record content rundown are inclined, with the help of latest advancements in content examination. Here, we present a multi-document synopsis, with the help of semantic likeness-based grouping, over the notable disseminated figuring structure of MapReduce.

The rest of the chapter is structured as follows. Section 2 elaborates related work for Big Data summarization; Section 3 examines the advancement used in innovation for Big Data; Section 4 proposes a framework for big information synopsis; and Section 5 traces test setup of the proposed information outline framework. This section gives an execution appraisal with the present calculations. The index of trial screenshots shows up in Section 6. Finally, Section 7 concludes the chapter.

6.2 Related Work

Gong and Liu generated rundown methodology utilizing latent semantic analysis (LSA). This technique detaches a group of words that make complete sense, with the best once-over a propelling power concerning the fundamental particular vector by LSA [11]. Zha utilized the normal fortress oversee **Mutual Reinforcement Principle** (MRP) and sentence gathering for the level once-over. This technique packs sentences of records into two or three topical get-togethers by sentence-gathering system. Groups of words that make complete sense are removed from each gathering, which have similar or related topics by deriving mathematical formulas which arrive at a score utilizing the MRP (i.e., changed LSA strategy) [12]. Yeh et al. proposed the design system utilizing LSA and the substance relationship portray. Semantic sentences utilizing LSA are found out by their system. Text relationship map [TRM] is made by the semantic sentences, and the essential sentences are expelled by the measure of relationship in TRM [13]. Li et al. broadened the dull multi-record graph utilizing LSA for the request based on conceptual report [14]. Han et al. put forward a proposal that delineated the substance rundown method utilizing criticism of Big Data (i.e., a request change process by part the concealed investigation into two or three pieces) [15]. Diaz and Gervas proposed a thing format framework for the personalization of news development structures. The technique utilizes three verbalization confirmation heuristics that grows once-overs utilizing two nonexclusive edited compositions and one changed summation relying on relevance feedback (RF) from news things [16]. In addition, they proposed an altered revamp rundown utilizing a mix of nonspecific and changed frameworks. Their nonspecific once-finished strategies join the position strategy with the topical word system. Their changed philosophy picks those sentences of a record that are most related to a given client [17]. Kurasova et al made altered graphs utilizing nonspecific and client particular strategies in light of likelihood. This framework expels the best arranging sentences by strategies for the nonexclusive sentence scoring and the client particular sentence scoring [18]. Ko et al. put forth a proposal that outlined from web pages utilizing Pseudo-Relevance Feedback (PRF) and a demand uneven summation in context of the likelihood display [19]. Li and Chen disconnected the altered substance bits, utilizing the likelihood movement examination and the secured Markov show [20]. Park et al. proposed the narrative summary frameworks, utilizing sentence arranging relying on the semantic highlights of the Mutual Reinforcement Principle (NMF) [21]. Nagwani proposed a framework for summarizing big text data [22]. Kolte and Bakal presented a framework for Big Data summarization technique, using new clustering and semantic similarity [23].

6.3 Technology for Big Data

Big Data is another idea for taking care of huge data. Consequently, the building portrayal of this innovation is exceptionally new [24]. Opensource structures like Apache Hadoop are putting away and preparing huge datasets utilizing clusters of commodity hardware. Hadoop is intended to scale up to hundreds and more nodes and is likewise profoundly blame tolerant. The different parts of a Hadoop stack are shown in Figure 6.1. The Hadoop stage contains accompanying essential parts.

6.3.1 Hadoop-distributed File System

To store large datasets reliably and to stream those datasets at high bandwidth to user applications, an open-source Hadoop-distributed file system (HDFS) framework is designed. Data in HDFS are recreated over different nodes for the figure execution and solid data security [25].

6.3.2 Data Processing

In the Hadoop system, the MapReduce and Hbase parts are utilized for information preparation. MapReduce is a parallel preparing structure which is a greatly versatile, parallel handling framework that works in pair with the

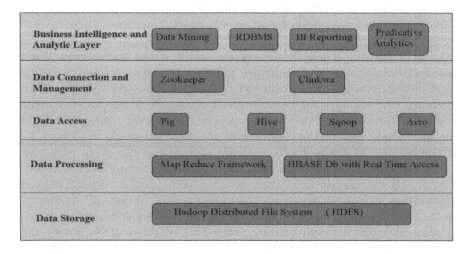

FIGURE 6.1
Hadoop framework

HDFS. A MapReduce work more often comprises three stages: map, copy, and reduce. The data are split into pieces of 64-MB size (as a matter of course). In the map phase, a client-characterized work acts on each chunk of data, delivering middle-of-the-road key-esteem sets, which are put away on neighborhoods plate. One map process is invoked to process one chunk of data. In the copy stage, the halfway key-esteem sets are exchanged to the area where a lighter procedure would work on the middle-of-the-road data. In reduce stage, a client-characterized decrease work acts on the middle-of-the-road key esteem that combines and produces the yield. One reduced process is invoked to process the scope of the keys.

HBase [26] is an open-source, conveyed, and nonsocial database framework actualized in Java. It keeps running over the layer of HDFS. It can serve the info and yield for the MapReduce in a very much organized structure. It is a Not Only Structure Query Language (NoSQL) database that keeps running over Hadoop as a disseminated and adaptable Big Data store. HBase enables you to question singular records and additionally infer total investigative reports on a huge measure of data [27].

6.3.3 Data Access

A specific arrangement of information being created for MapReduce is called wrapper. These wrappers can give a superior control over the MapReduce code and help in the source code improvement. The accompanying wrappers are broadly utilized as a part of blend with MapReduce. Pig [28] is a high-level state stage, where the MapReduce structure is made, which is utilized with Hadoop stage. It is a high-level state information handling framework, where the information records are dissected in a high-level state dialect. Pig is a high-level state stage for making MapReduce programs work in hand with Hadoop, and the dialect we use for this stage is called Pig Latin. The Pig was intended to make Hadoop more easy to understand and congenial to control by clients and non-designers. Pig is an intuitive execution condition supporting Pig Latin dialect. Hive [29] is an application created for information stockroom that gives the Structure Query Language (SQL) interface and additionally a social model. Hive framework is based on the best layer of Hadoop that assists in giving conclusions and investigation results for individual inquiries [30]. Sqoop [31] is a command-line interface application that gives stage which is accustomed to changing over information from social databases and Hadoop or the other way around. It can be utilized to populate tables in Hive and Hbase. Sqoop utilizes a connector-based engineering which underpins modules that give availability to outside frameworks. Sqoop incorporates connectors for databases, for example, MySQL, Postgre SQL, Oracle, SQL Server, and DB2 and nonexclusive JDBC connector. Dataset being exchanged is cut into various portions and a guide for employment is propelled with singular mappers that is in charge of exchanging a portion of this dataset. Sqoop utilizes the database metadata to derive information it composes. Avro [32] is a framework that gives usefulness of data serialization and administration of data trade. It is fundamentally utilized as a part of Apache Hadoop. These administrations can be utilized together and in addition freely agree with the data records. Avro depends on diagrams. At the point when Avro data are perused, the pattern utilized when composing it is constantly present. This encourages use with dynamic, scripting dialects, since information, together with its pattern, is completely self-depicting. At the point when Avro data are put away in a document, its construction is put away with it; so, records might be handled later by any program.

6.3.4 Data Association and Administration

Zookeeper and Chukwa are utilized for data association and administration to process and break down the gigantic measure of logs.

Zookeeper [33] is an open-source Apache venture that brings together framework and administrations that empower synchronization over a group. Zookeeper keeps up basic articles required in substantial bunches. Chukwa [34] is an open-source data accumulation framework for observing substantial disseminated frameworks. It is based over the HDFS and MapReduce structure and acquires Hadoop's adaptability and power. It additionally incorporates a flexible and effective toolbox for showing, observing, and examining results to make the best utilization of the gathered data.

6.4 Proposed Summarization Framework

Figure 6.2 shows the architecture of the proposed system. The working of the proposed data summarization is described below in detail.

6.4.1 Preprocessing Data

Preprocessing of data is necessary to make data equipped for processing. Most of the real time data is incomplete, redundant, and uncertain. Thus using preprocessing, we can get clean data. In preprocessing of the data, we use data cleansing, data integration, data transformation, data reduction, and data compression.

6.4.2 New Method of Data Clustering

Despite many clustering algorithms that are available, the current scenario requires a highly scalable clustering algorithm. To improve scalability, the algorithm must work with all types of data (numerical, categorical, or binary). Clustering algorithm must be able to find out clusters of arbitrary shapes. It should not work only with distance measures that are capable of finding out spherical clusters of small size. The clustering algorithm should be able to handle both low as well as high-dimensional data.

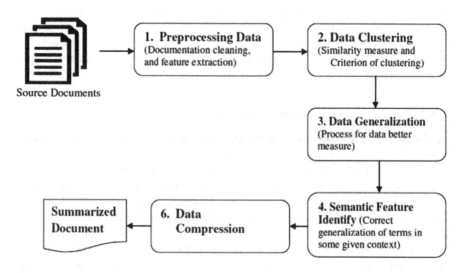

FIGURE 6.2
Framework of Big Data summarization

6.4.3 Modified Fuzzy C-means Clustering

Fuzzy C-means (FCM) algorithm, also called fuzzy Iterative Self-Organizing Data (ISODATA) was presented by Bezdeck [35] as an augmentation to Dunn's calculation [36]. The FCM-based algorithm is the most generally utilized fuzzy clustering algorithm [37-39]. However, in FCM, there are a few imperatives that influence the execution [40-43]. The main impediment is the determination of irregular centroids at introductory level. Thus, the calculation sets aside greater opportunity to discover groups. The second limitation is its failure to compute the enrollment esteem if separations of information point are zero, while the proposed modified fuzzy C-means (MFCM) algorithm at first figures centroids suitably and proposes new part capacity to ascertain the participation esteem, regardless of whether separations of information point are zero.

Let $X = \{x_1, x_2, \ldots, x_n\}$, where $x_i \in \Re^n$ present a given set of feature data. Minimizing cost function is the main objective of MFCM algorithm.

$$J(U, V) = \sum_{j=1}^{C} \sum_{i=1}^{C} \left(\mu_{ij}\right)^m \|x_i - v_j\|^2 \tag{1}$$

$V = \{v_1, v_2, \ldots, v_c\}$ are the cluster centers. The cluster centers are initially calculated as follows. To determine the centroid of the cluster, all the patterns are applied to every other pattern and the patterns having Euclidian distance less than or equal to α (user-defined value) are counted for all the patterns. Later, the pattern with the maximum count is selected as the centroid of the cluster. If

$$\left(\left|R_i \frac{p}{\underset{j=1}{}} R_j\right| \leq \alpha\right) \tag{2}$$

$$\text{then } D_i = D_i + 1, \quad \text{where } i = 1, 2, \ldots, p.$$

If $D_{\max}$ is the maximum value in the row vector D and D_{ind} is the index of maximum value, then

$$[D_{\max} D_{\mathrm{ind}}] = \max[D]; \quad C_1 = R_{\mathrm{ind}}$$

$U = (\mu_{ij})_{N \times C}$ is the fuzzy partition matrix, in which each member μ_{ij} indicates the degree membership between the data vector x_i and cluster j. The values of the matrix U should satisfy the following conditions:

$$\mu \in [0, 1], \quad \forall i = 1, \ldots, N, \quad \forall j = 1, \ldots, C \tag{3}$$

$$\sum_{j=1}^{C} \mu i_j = 1, \quad \forall i = 1, \ldots, N \tag{4}$$

Appropriately initialize the membership matrix U using

$$f(x, v, r) = 1 - f(\cdot),$$

$$where\ f(\cdot) = \begin{cases} 1 & \text{if } r \geq 1 \\ 0 & \text{if } r = 0 \\ r\gamma & \text{if } 0 < r < 1 \quad (\gamma \text{ is sensitive parameter}) \end{cases} \tag{5}$$

$r = \left\| x_i - v_j \right\|$, $r \geq 1$, and if $r > 1$, then $r\gamma$ is set to 1.

To satisfy conditions 2 and 3, divide the total sum of attributes to the each attribute for every pattern.

The exponent $m \in [1, \infty]$ is the weighting exponent which determines the fuzziness of the clusters. Minimization of the cost function $J[U, V]$ is a nonlinear optimization problem, which can be minimized with the following iterative algorithm:

Step 1: Find the appropriate centroids using equation (2).
        Choose appropriate exponent m and termination criteria.
Step 2: Initialize the membership matrix U using equation (4).
Step 3: Calculate the cluster center V according to the equation:

$$v_j = \frac{\sum_{i=1}^{N} \left(\mu_{ij} \right)^m x_{ij}}{\sum_{i=1}^{N} \left(\mu_{ij} \right)^m}, \quad \forall j = 1, \ldots, C \tag{6}$$

Step 4: Calculate new distances norm:

$$r = \left\| x_i - v_j \right\|; \quad \forall i = 1, \ldots, N, \quad \forall j = 1, \ldots, C$$

Step 5: Update the fuzzy partition matrix U:

If $r > 0$ (indicating that $(x_i \neq v_j)$)

6.4.4 Data Generalization

Data generalization is the way toward making next layers of rundown information. Latent Dirichlet allocation (LDA) is a prevalent theme-displaying system, which models the content reports as blends of inactive points that are initial ideas exhibited in the content. A subject model is a likelihood dissemination procedure over the accumulation of content archives, where each record is displayed as a mix of themes, which speaks to clusters of words that have a tendency to happen together. Every theme is demonstrated as a likelihood conveyance φ over lexical terms. Every subject is displayed as a vector of terms with the likelihood in the vicinity of 0 and 1. An archive is demonstrated as a likelihood circulation over themes. In LDA, the point blend is drawn from a conjugate Dirichlet, which is the same for all records. The point displaying for content gathering utilizing LDA is performed in four stages. In the initial step, multinomial θt dissemination for every subject t is chosen from a Dirichlet circulation with parameter β. In the second step, for each report d, a multinomial appropriation θb is chosen from a Dirichlet conveyance with parameter α. In the third step, for each word w in record s, a subject t from θb is chosen. Finally, a word w from θt is chosen to speak to the subject for the content report. The likelihood of producing a corpus is given by the following condition:

$$p(D|\alpha,\beta) = \prod_{d=1}^{M} \int p(\phi_d|\alpha) \left(\sum_{n-1}^{N_d} p(Z_{dn}|\phi_d)p(w_{dn}|\phi_d)p(\theta|\beta) \right) d_{\vartheta d}d_p$$

The above condition helps in generalizing the data in a series of steps that modify pieces of important information into low and high categories depending on their similarity levels. In this step, LDA-point-demonstrating system is applied on every individual substance report, to make grouped subjects and terms having a place in each theme.

6.4.5 Process of identifying similar terms

Grouping the terms semantically provides information in a clear way, thereby allowing its optimization for further learning, exposure, and understanding of information. In this step, semantic relative terms are set up for each subject term conveyed in the previous stage. WordNet Java API [45] is utilized to convey the outline of semantic close terms. The sentence whose meaning is conveyed by a group of words (close terms) is conveyed over the MapReduce structure and the configured semantic terms are added to the vector. The Mapper and Reducer for semantic term generation from each cluster are given as follows:

//Mapper:

1. For each keyword term in keyword list $\{K_1, K_2, ..., K_n\}$
2. Get the semantic similar term KS_i = ComputeSematicSimilar(K_i) (Get ahead of the term K_i in wordnet API and take out similar term in the set KS_i)

3. For all keyword n, $n \in KS_i$ present all documents D do.

//Reducer:

1. For each keyword K, count $\{C_1, C_2, ..., C_n\}$
2. First set the sum term of keyword occurrence as 0
3. For all count $C \in$ count $\{C_1, C_2, ..., C_n\}$ do
4. Revise count sum = sum + C
5. Count sum

6.4.6 Data Compression

The guideline to follow for this algorithm is to divide the information into two parts: first, the essential information will contain interesting nonzero bytes, and the second information will contain bits in regard to the elucidating position of nonzero and zero bytes. The two parts of information by then can be packaged with other information compression algorithms to achieve the most compressed proportion. The compression method can be depicted as below:

//Mapper:

1. Read input from source file byte by byte
2. Classify the bytes read from the source file as zero or nonzero byte
3. Create a temp byte file and mark bit "1" for nonzero byte data and bit "0" for zero byte data
4. Follow the above three steps till temporary byte file reaches 8 bits
5. When temp byte data are filled with 8 bits, byte value of this is written into next data file
6. Temp byte data are cleared
7. Follow reading the data mentioned above in the six steps until the end of source file

/Reducer:

1. Combine the data and write into a fresh file by specifying
   a. Original i/p data length
   b. Temp byte file
   c. Next data file

6.5 Experiment evaluation

6.5.1 Performance Evaluation of Clustering

We have done an experimental analysis with the proposed procedure, *K*-means, FCM, and *K*-medoids to make three groups without knowing class information. In *K*-means, centroid selection is done randomly. Yet, various distinctive choices are open including Al-Daoud and Roberts [46], Khan and Ahmad [47], and others. Table 6.1 depicts the outcome obtained from the clustering system. It shows that the proposed grouping count gives superior results compared to the ordinary estimations. The computation determines centroids suitably instead of picking arbitrary values.

Assessment of data summarization: The process of summarization of data is assessed by precision, recall, and computational time.

Precision: The precision is defined as the ratio of relevant documents selected to number of documents selected. The precision represents the probability that a selected item is relevant.

$$P = \frac{N_{rs}}{N_s}$$

Recall: The recall is defined as the ratio of relevant documents selected to total number of relevant documents available. The recall is a measure of completeness.

$$R = \frac{N_{rs}}{N_r}$$

The implementation of Big Data text summarization is carried out using the Java-based open-source technologies and the MapReduce. The dataset

TABLE 6.1

Comparative analysis of clustering algorithms

Algorithms	Clustering performance for each set		
	Set 1 (%)	Set 2 (%)	Set 3 (%)
K-means	100	63	52
K-medodis	100	67	64
Fuzzy C-means	100	74	73
Modified fuzzy C-means	100	89	87

TABLE 6.2

Comparison of proposed and existing system

Algorithms	Precision	Recall (%)	Time (s)
LSA	0.65	0.63	55
BDS	0.69	0.64	45
BDS-FSCM	0.81	0.75	15

is available on the Internet. The experiments are performed on Windows 7 operating system with four nodes. The precision, recall, and computational times are used for measuring the quality of data summarization. The comparative results of the proposed and existing systems, LSA [48] and Big Data Summarization (BDS) [49], are tabulated in Table 6.2.

6.6 Conclusion

This chapter presents a novel approach for fuzzy clustering, semantic approach, and data compression for Big Data summarization. K-means algorithm is mostly used in traditional summarizations; however, it performs poor due to several user-specified input parameters. Our proposed modified FCM can solve this problem up to some extent. The outcome from different recreations utilizing Iris informational index demonstrates that the proposed modified FCM performs superior to K-means, K-medoids, and FCM grouping, which enhances the quality information outline. The proposed semantic approach and data compression techniques enhance the superiority of data summarization. Precision, recall, and computational time of the proposed framework are superior to LSA and BDS.

References

[1] Schmidt, Data is exploding: the 3 V's of. Business Computing World, 2012.
[2] Y. Zhai, Y.-S. Ong, and IW. Tsang, The Emerging "Big Dimensionality". In Proceedings of the 22nd International Conference on World Wide Web Companion, Computational Intelligence Magazine, IEEE, vol. 9, no. 3, pp. 14–26, 2014.
[3] V. Medvedev, G. Dzemyda, O. Kurasova, and V. Marcinkevičius, "Efficient data projection for visual analysis of large data sets using neural networks", Informatica, vol.22, no. 4, pp. 507–520, 2011.
[4] G. Dzemyda, O. Kurasova, and V. Medvedev, "Dimension reduction and data visualization using neural networks", in Maglogiannis, I., Karpouzis, K., Wallace, M., Soldatos, J., eds.: Emerging Artificial Intelligence Applications

in Computer Engineering. Volume 160 of Frontiers in Artificial Intelligence and Applications, IOS Press, 2007, pp. 25–49.

[5] McCallum, K. Nigam,and L. Ungar, "Efficient clustering of high-dimensional data sets with application to reference matching", in Proceedings of the 6th ACM SIGKDD International Conference on Knowledge Discovery and Data Mining, pp. 169–178, 2000.

[6] M.H. Dunham, Data Mining: Introductory and Advanced Topics, Prentice Hall PTR, Upper Saddle River, NJ, USA, 2002.

[7] MacQueen, "Some methods for classification and analysis of multivariate observations", in Le Cam, L.M., Neyman, J., eds.: Proceedings of the Fifth Berkeley Symposium on Mathematical Statistics and Probability, Berkeley and Los Angeles, CA, USA, University of California Press, vol. 1, pp. 281–297, 1967.

[8] T. Kohonen, Overture. Self-Organizing neural networks: recent advances and applications, Springer-Verlag, New York, NY, USA, 2002.

[9] Turpin A., Tsegay Y., Hawking D., & Williams H. (2007) Fast generation of result snippets inweb search. In Proceedings of the 30th annual international ACM SIGIR conference on Researchand development in information retrieval, Amsterdam, Canada.

[10] Sampath G., & Martinovic M. (2002) A Multilevel Text Processing Model of Newsgroup Dynamics. In Proceedings of the 6th International Conference on Applications of Natural Language to Information Systems, NLDB 2002.

[11] Y. Gong, X. Liu, "Generic Text Summarization Using Relevance Measure and Latent Semantic Analysis", in Proceedings of the 24th annual international 1102 Yoo-Kang Ji et al. ACM SIGIR conference on research and development in information retrieval (SIGIR'01), pp. 19-25, New Orleans, USA, 2001.

[12] H. Zha, "Generic Summarization and Keyphrase Extraction Using Mutual Reinforcement Principle and Sentence Clustering", In Proceedings of the 25th annual international ACM SIGIR conference on research and development in information retrieval (SIGIR'02), pp. 113-120, Tampere, Finland, 2002.

[13] J. Y. Yeh, H. R. Ke, W. P. Yang, I. H. Meng, "Text summarization using a trainable summarizer and latent semantic analysis", Information Processing and Management 41, pp. 75-95, 2005.

[14] W. Li, B. Li, M. Wu, "Query Focus Guided Selection Strategy for DUC 2006", In Proceedings of the Document Understanding Conference (DUC'06), 2006

[15] K S Han, D H Bea, and H C Rim, "Automatic Text Summarization Based on Relevance Feedback with Query Splitting," In Proceedings of the 5th International Workshop on Information Retrieval with Asian Language, Hong Kong, pp. 201-2, Sep. 2000.

[16] A Diaz, and P Gervas, "Item Summarization in Personalisation of News Delivery Systems," In Proceeding of the 7th International Conference on Text, Speech and Dialogue (TSD), LNAI 3206, Brno, Czech Republic, pp. 49-56, Sep. 2004.

[17] Yoo-Kang Ji, Yong-Il Kim, Sun Park. "Big data summarization using semantic feture for IoT on cloud", Contemporary Engineering Sciences, 2014.

[18] Kurasova, Olga, Virginijus Marcinkevicius, Viktor Medvedev, Aurimas Rapecka, and Pavel Stefanovic. "Strategies for Big Data Clustering", 2014 IEEE 26th International Conference on Tools with Artificial Intelligence, 2014.

[19] Y J Ko, H K An, and J Y Seo, "Pseudo-relevance feedback and statistical query expansion for web snippet generation," Information Processing Letters, vol. 109, pp. 18-22, 2008.

[20] Q Li, and Y P Chen, "Personalized text snippet extraction using statistical language models," Pattern Recognition, vol. 43, pp. 378-86, 2010.

[21] J H Lee, S Park, C M Ahn, and D H Kim, "Automatic Generic Document Summarization Based on Non-negative Matrix Factorization," Information Processing and Management, vol. 45, pp. 20-34, Jan. 2009.

[22] N K Nagwani, "Summarizing Large Text Collection using Topic Modeling and clustering based on MapReduce Framework", Journal of Big Data, Springer, 2014.

[23] Shilpa Kolte and J W Bakal, "Big Data Summarization Using Novel Clustering Algorithm And Semantic Feature Approach", International Journal of Rough Sets and Data Analysis, IGI-Global publication, USA, Vol. 4, Issue 3, 2017.

[24] Bhatt, Chintan, Nilanjan Dey, and Amira S. Ashour, eds. "Internet of things and big data technologies for next generation healthcare." (2017): 978-3.

[25] Tamane, Sharvari, Sharvari Tamane, Vijender Kumar Solanki, and Nilanjan Dey. "Privacy and security policies in big data." (2017).

[26] K V N Rajesh. Big Data Analytics: Applications and Benefits, 2009.

[27] Dey, Nilanjan, Amira S. Ashour, and Chintan Bhatt. "Internet of things driven connected healthcare." In Internet of things and big data technologies for next generation healthcare, pp. 3-12. Springer, 2017.

[28] Kamal, Md Sarwar, Sazia Parvin, Amira S. Ashour, Fuqian Shi, and Nilanjan Dey. "De-Bruijn graph with MapReduce framework towards metagenomic data classification." International Journal of Information Technology 9, no. 1 (2017): 59-75.

[29] Dey, Nilanjan, Aboul Ella Hassanien, Chintan Bhatt, Amira Ashour, and Suresh Chandra Satapathy, eds. Internet of Things and Big Data Analytics Toward Next-Generation Intelligence. Springer, 2018.

[30] Kamal, Sarwar, Shamim Hasnat Ripon, Nilanjan Dey, Amira S. Ashour, and V. Santhi. "A Map Reduce approach to diminish imbalance parameters for big deoxyribonucleic acid dataset." Computer methods and programs in biomedicine 131 (2016): 191-206.

[31] Kamal, S., N. Dey, A. S. Ashour, S. Ripon, V. E. Balas, and M. S. Kaysar. "Fb Mapping: An automated system for monitoring Facebook data." Neural Network World 27, no. 1 (2017): 27.

[32] J. Baker, C. Bond, J. Corbett, J. Furman, Lloyd, and V. Yushprakh. Megastore: providing scalable, highly available storage for interactive services. In Proceedings of Conference on Innovative Data Systems Research, 2011.

[33] https://www.tutorialspoint.com/zookeeper/zookeeper_tutorial.pdf

[34] J. Boulon, A. Konwinski, R. Qi, A. Rabkin, E. Yang, and M. Yang. Chukwa, A large-scale monitoring system. In First Workshop on Cloud Computing and its Applications (CCA '08), Chicago, 2008.

[35] J. Bezdek. "Pattern Recognition with Fuzzy Objective Function Algorithms". Plenum Press, USA, 1981.

[36] J.C. Dunn. "A Fuzzy Relative of the ISODATA Process and its Use in Detecting Compact, Well Separated Clusters". Journal of Cybernetics, 3(3): 32-57, 1974.

[37] Rani, Jyotsna, Ram Kumar, F. Talukdar, and Nilanjan Dey. "The brain tumor segmentation using fuzzy c-means technique: a study." Recent advances in applied thermal imaging for industrial applications, 40-61, 2017.

[38] Wang, Dan, Zairan Li, Nilanjan Dey, Amira S. Ashour, R. Simon Sherratt, and Fuqian Shi. "Case-based reasoning for product style construction and fuzzy analytic hierarchy process evaluation modeling using consumers linguistic variables." IEEE Access 5 (2017): 4900-4912.

[39] Ahmed, Sk Saddam, Nilanjan Dey, Amira S. Ashour, Dimitra Sifaki-Pistolla, Dana Bălas-Timar, Valentina E. Balas, and João Manuel RS Tavares. "Effect of fuzzy partitioning in Crohn's disease classification: a neuro-fuzzy-based approach." Medical & biological engineering & computing 55, no. 1, 2017.

[40] Ngan, Tran Thi, Tran Manh Tuan, Nguyen Hai Minh, and Nilanjan Dey. "Decision making based on fuzzy aggregation operators for medical diagnosis from dental X-ray images." Journal of medical systems 40, no. 12, 2016.

[41] Wang, Dan, Ting He, Zairan Li, Luying Cao, Nilanjan Dey, Amira S. Ashour, Valentina E. Balas et al. "Image feature-based affective retrieval employing improved parameter and structure identification of adaptive neuro-fuzzy inference system." Neural Computing and Applications 29, no. 4, 2018.

[42] Tuan, Tran Manh, Hamido Fujita, Nilanjan Dey, Amira S. Ashour, Vo Truong Nhu Ngoc, and Dinh-Toi Chu. "Dental diagnosis from X-Ray images: An expert system based on fuzzy computing." Biomedical Signal Processing and Control pp. 64-73, 2018.

[43] Wang, Cunlei, Zairan Li, Nilanjan Dey, Amira Ashour, Simon Fong, R. Simon Sherratt, Lijun Wu, and Fuqian Shi. "Histogram of oriented gradient based plantar pressure image feature extraction and classification employing fuzzy support vector machine." Journal of Medical Imaging and Health Informatics, 2017.

[44] Karaa, Wahiba, Ben Abdessalem, and Nilanjan Dey. Mining Multimedia Documents, CRC Press, 2017.

[45] WordNet Java API.

[46] AI-Daoud, M. B., & Roberts, S. A. (1996). New methods for the initialization of clusters. Pattern Recognition Letters, 17, 451–455.

[47] Khan, S. S., & Ahmad, A., (2004). Cluster center initialization algorithm for K-means clustering. Pattern Recognition Letters, 25, 1293–1302.

[48] Saba, Luca, Nilanjan Dey, Amira S. Ashour, Sourav Samanta, Siddhartha, and Jasjit S. Suri (2016) "Automated stratification of liver disease in ultrasound: an online accurate feature classification paradigm." Computer methods and programs in biomedicine pp. 118-134.

[49] Y. Gong, X. Liu (2001) Generic Text Summarization Using Relevance Measure and Latent Semantic Analysis. Proceedings of the 24th annual international ACM SIGIR conference on Research and development in information retrieval, New Orleans, Louisiana, United States.

7

Topic-Specific Natural Language Chatbot as General Advisor for College

Varun Patil, Yogeshwar Chaudhari, Harsh Rohila, Pranav Bhosale, and P. S. Desai
Department of Computer Engineering, Smt. Kashibai Navale College of Engineering, Pune

CONTENTS

7.1 Introduction

A **chatbot** (also known as an **interactive agent** or **artificial conversational entity**) is a computer program which conducts a conversation via auditory or textual methods. Such programs are designed to replicate the role of a human conversational partner. Naturally, chatbot's application can be extended in daily life, such as help-desk tools, automatic telephone-answering systems, and tools to aid in education, business, and e-commerce [1, 2]. The emerging trend in online chatting is gaining popularity, where a customer can chat with customer care representatives at any point of time from anywhere [3]. We are looking at designing a system to make natural conversations between human and machine that is specifically aimed at providing a chatbot system for members of a particular organization. Proposed system uses natural language processing to provide an appropriate response to the end user for a requested query. The bot is based on natural language; so, a user does not have to follow any specific format. However, some queries might start with special symbols to be used for special intents, for example, if the user wants Google search results for a topic, then the query will be appended by a designated special symbol like "#."

The system will use natural language processing (using Dialogflow formerly known as Api.ai) to understand the user-provided query. Being a topic-specific system, user can query about any topic (here college)-related events using June (chatbot). This eliminates the hassle of visiting college for trivial queries. After analyzing the user's query, system will strive to generate the response that will satisfy the query. The system administrator can update the existing model to handle more diverse queries. Figure 7.1 describes various fundamental use cases that the user or administrator may use. Viewing information is an important use case, which has been focused in our system.

By building an automated intelligent system to resolve the domain-related queries of user, many human resources previously used for resolving queries now can be utilized for other purposes, which will increase productivity. Moreover, the system's availability, response time, and ability to handle a number of clients simultaneously make it more suitable to handle this task. After analyzing multiple methods for building chatbots, it was observed that most of the architectures depend on natural language understanding than pattern matching. It also helps in understanding many diverse queries. The system consists of a Controller which is a combination of the knowledge base and the query processing engine, and a View which is a simple interface for system–user interaction which enables user to learn the system quickly and use it efficiently.

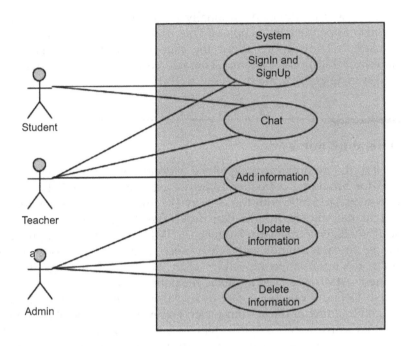

FIGURE 7.1
Use case diagram (major use cases explained).

7.2 Problem Definition

The objective of the system is to implement an automated system to simulate a domain-specific intelligent conversation between user and computer to deliver the exact piece of data depending on a query.

We aim to design and develop a system that will be capable of holding natural, context-aware conversations with an end user to provide a response to the associated query. This system will be designed to resolve the queries associated with the given domain/organization based on the available knowledge. The system should provide a way to update and add new information to the domain knowledge base. The system will use natural language processing to identify the intent of the user's query. Training data consisting of sample queries will be used to train the system for possible queries that can be asked by the user. If the system is unable to process a query using the available knowledge base and training dataset, then the response for the query will be generated based on the corresponding Google search. Based on the intent of the query, conversation context and knowledge-base response formulation will be done. Additionally with actual data, extra information is also appended in response. This makes it more humanized and helps user to know more about the data that would not have been possible in a traditional system. Generated response is then provided to the end user to resolve the query.

7.3 Literature Survey

In Ref. [4], the authors have explained a design of a chatbot, specifically tailored for providing a frequently answered question (FAQ) bot system for university students, with the objective of mimicking an undergraduate advisor at the student information desk. The design semantics includes artificial intelligence mark-up language (AIML) specification language for authoring the information repository such that the chat robot design separates the information repository from the natural language interface component. AIML was used to define the patterns of the questions that can be asked. Defining question patterns in AIML is very simple that makes the overall building of the system easier. For every new type of question to be handled, corresponding AIML pattern needs to be added. This makes updating question-sets tedious and too difficult to manage. Moreover, it is very difficult to develop and maintain a contextual understanding when pattern matching is being used for extracting information from queries. In Ref. [5], design of a college enquiry chatbot is explained. This chatbot is built using artificial algorithms that analyze user's queries and understand user's messages. In Ref. [6], an automated question-answering system is introduced. The system answers the queries posted by the student in a more interactive way like a virtual teacher (chatbot system). Naive Bayesian Classifier is used to identify the question being asked. Although this classifier is very accurate for classification, the amount of questions that can be classified is limited for such classifiers. As the number of questions increases, its accuracy of classifying questions will decrease relatively. In Ref. [7], a chatbot is proposed which automatically gives immediate response to the users based on the dataset of FAQs, using AIML and latent semantic analysis. Semantic analysis will handle the shortcomings of AIML. AIML is incapable of handling the contextual information, but by incorporating semantic analysis, the task of maintaining contextual information will become easier. However, the problem of defining large number of patterns for handling various types of questions still persists with AIML. In Ref. [8], an android application is proposed for visually impaired people. It gives an answer to the education-based queries asked by the visually impaired people. Once the application is open, it gives a voice instruction to use an application. The output will be provided in voice form as well as in text form. In Ref. [9], the authors have proposed an idea of building a chatbot based on ontological approach. It maps the domain-related knowledge and information in a relational database. This relational database is then utilized to provide relevant information which can be used to generate chatbot responses. This approach is suitable for developing a chatbot for domain-specific requirements, as a domain-specific chat. The system architecture proposed in this work is influenced by the ontological approach.

7.4 System Architecture

The architecture of the system is divided into three parts. As shown in Figure 7.2, first is the user interface (Android app or web browser) using which the user can fire a query. Next, we have the "June Core" which implements all the processing required for generating a response. At last, we have Api.ai service which is used for providing natural language-processing capabilities.

The user interacts with the system using the user interface provided in the system. The interface will have the look of a typical chat interface where a text area is provided to enter the query. The front-end performs preprocessing on the query (if required) and the preprocessed query will be forwarded to the system's back-end. The user interface will also manage the contextual information which can be utilized for generating a user specific response.

The system's backend utilizes the Api.ai service, provided by Google. Api.ai will act as a natural language-processing engine for the system core. Simple domain irrelevant queries such as "Hi!!" and "How are you?" will be handled by Api.ai module's "small talks." Although domain-related queries are handled by training the system with a set of sample queries to build the knowledge base, this knowledge base will be utilized to identify an intent of a query and entity keywords in a query. The result will be returned in the form of JavaScript Object Notation object which has fields like intent, entities, action, metadata, and custom payload data such as queryType, and so on and other information related to the query. This JSON object will be used for query formulation in further system processing and then will be forwarded to the webhook where the logic to generate the query response resides. June system processes the JSON object returned by Api.ai. Using the JSON object, the system will formulate the database query based on the values of the entities, actions, and by combining the intent parameters. The formulated database query will fetch the required information from the knowledge database (Knowledge_db).

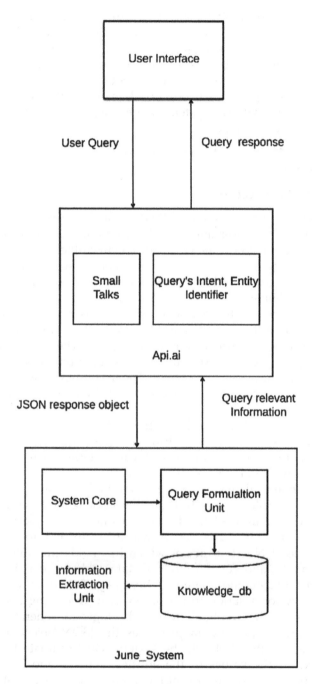

FIGURE 7.2
System architecture (brief system architecture with major components).

This response is processed to represent it in natural language. This response is sent to the user interface.

When the user query is passed to our system, each component processes the query and changes it from a message to a different data format along with some metadata. The message from the user gets converted into a JSON object. Furthermore, the backend cloud function processes the JSON object, and an appropriate query is generated which is forwarded to the database. Results from the database are again transformed into natural language response and forwarded to the user. This flow of data is given in Figures 7.3–7.5, where Figures 7.4 and 7.5 provide more detailed information regarding the movement of data.

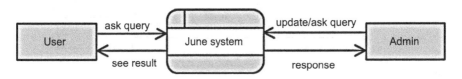

FIGURE 7.3
DFD_level_1 diagram (context of the proposed system).

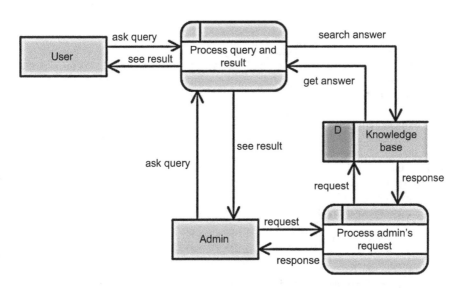

FIGURE 7.4
DFD_level_2 diagram (breakdown of main activities in proposed system).

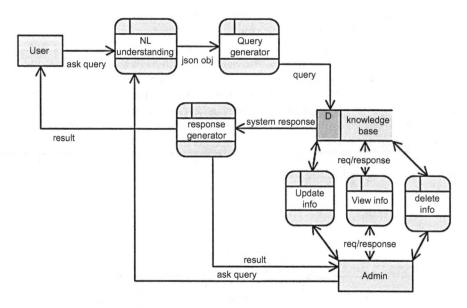

FIGURE 7.5
DFD_level_3 diagram (detailed analysis of activities in proposed system).

7.5 Algorithm

```
Accept user query
Query = query + userMetaData;
sendQuery( What is my attendance in second semester?
+userMetaData)
While(prompt)
{
     Wait for user response
     sendQuery( Missing parameter + userMetaData);
}
Call action
Filter(parameters)

if (para1 == dummy )
{
    switch(para7)
    {
    case case 1 :
    if (para2 =+ metaEntityValue && para3 == sample
    EntityValue)
```

```
                {
                        Construct query with received data & avail-
                        able data. Fetch result from database.
                        Construct result. sendResponse()
                } else
                {
                        Execute else.
                }
                .
                .
                .
                }
                }
                .
                .
                if(resultConflict)
                {
                Wait for user selection
                Accepted input : 1, first, second one, later one,
                last one. Parse response for selected option and
                display that only. (happens locally)
                } else
                {
                        Display response.append(Data insights if
                        available);
                        Wait for new request.
                        }
        }
}
```

7.6 Implementation

This section describes the implementation for developing the chatbot. A Dialogflow agent is created and trained to understand user queries. A web server is implemented to generate response and an android app is developed for user interface.

7.6.1 Android App

7.6.1.1 Technologies Used

Java
Android Studio
Cloud Firestore and Firebase Auth

The app mainly consists of three activities/pages: chat activity, sign-in activity, and sign-up activity.

7.6.1.2 Chat Activity

This is the entry point of the app. The user will be able to ask queries here. This activity will contain options to sign-in page and sign-up page.

7.6.1.3 Sign-Up Activity

This provides sign-up functionality for two types of users, that is, student and teacher. Cloud Firestore is used for storing user details. User email is used as a unique id.

7.6.1.4 Sign-In Activity

Figure 7.6 depicts that the system provides sign in using email and password for both type of users. Firebase Auth is used for authenticating users with email and password.

7.6.2 Dialogflow Agent

Some key terms are required to understand this part.

7.6.2.1 Entities

These are the things or objects that our chatbot agents need to understand user queries. For example, person name is an entity. Using this, the chatbot can understand that words like Sam, Ben, and others are person names. Similarly "color" is also an entity by which the chatbot can understand that words like blue, red, and others are colors.

7.6.2.2 Intents

Intents are used to understand the user's intention, that is, what information user wants to know. For example, if the user wants to know about a person, it can create person details intent.

Dialogflow agent is trained using entities and intents, for example, if user asks, "get me the email of Ben." To understand what the user wants, it can create a personInfo entity which can identify the word "email" as personInfo. Another entity can be a personName, which identifies the word "Ben" as personName. An intent getPersonDetails can be created which is returned by the agent if it finds personDetails and personName in one query.

The agent is trained with many ways in which the user can ask queries. For example, another way a user can ask the same question is, "email of Ben." With more examples, the agent can identify intents with more confidence.

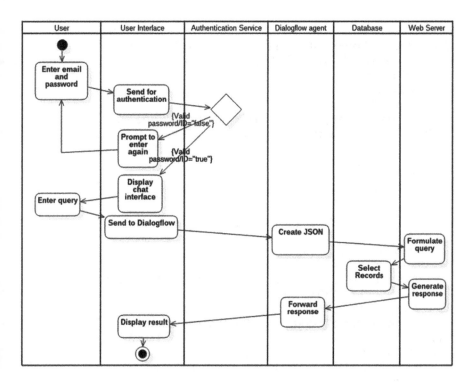

FIGURE 7.6
Activity diagram (dynamic flow control in the proposed system).

7.6.2.3 Actions

An action corresponds to the step of application that the agent takes when a specific intent has been triggered by a user's input. Actions can have parameters for extracting information from user requests and will appear in the following format in a JSON response:

```
{"action":"action_name"}
{"parameter_name":"parameter_value"}
```

7.6.2.4 Parameters

Parameters are elements generally used to connect words in a user's response to entities. In JSON responses to a query, parameters are returned in the following format:

```
{"parameter_name":"parameter_value"}
```

Here's an example of JSON that will be generated after query analysis. Response payload:

```
{
"id": "cf49a7e9-ed58-4344-874c-07ab974bbb2c",
  "timestamp": "2017-12-15T14:52:48.558Z",
  "lang": "en",
  "result": {
    "source": "agent",
    "resolvedQuery": "What is pranav's Number ? ",
    "action": "lookForPhone",
    "actionIncomplete": false,
    "parameters": {
      "contactAttribute": [
        "phone"
      ],
      "teacher": "Pranav"
    },
    "contexts": [],
    "metadata": {
      "intentId": "735166a4-70da-4b2b-8d37-dc7f460cf771",
      "webhookUsed": "false",
      "webhookForSlotFillingUsed": "false",
      "intentName": "teacherContactDetails"
    },
    "fulfillment": {
      "speech": "Pranav 's phone number is",
      "messages": [
        {
          "type": 0,
          "speech": "Pranav 's phone number is"
        },
        {
          "type": 4,
          "payload": {
          "queryType": "search",
          "requiredAttribute": "phone"
          }
        }
      ]
    },
    "score": 0.9599999785423279
  },
  "status": {
    "code": 200,
    "errorType": "success",
    "webhookTimedOut": false
  },
  "sessionId": "12bb615c-8d1d-b176-976d-c49cc1b94183"
}
```

7.6.3 Web Server

7.6.3.1 Technologies Used

Node.js

Google Cloud Functions Cloud Firestore

Figure 7.7 represents the flow of control from the various technologies used in the system. It starts from the user interface provided using the android application to the dialog flow wherein dialog flow forwards requests to the web server after processing. A Google Cloud function is created and deployed on Google Cloud. Node.js is used for cloud function. Cloud Firestore is used as a knowledge base. It stores data in collections, as a key-value format.

Web server processes the intent and entities returned by Dialogflow agent, to generate meaningful response. Intents are used to identify collections to query. Entities are used to formulate queries.

For example, intent is getPersonDetails, and entities are personDetails and personName.

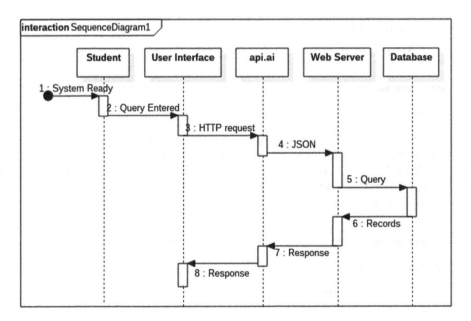

FIGURE 7.7
Sequence diagram (flow of control in proposed system).

7.6.4 Other Specification

7.6.4.1 Advantages

This system will serve as a point of contact between the administrator and a general user for most of the topic-related general queries.

It will reduce the frequency of trivial queries being asked to the administrator while maximizing number of customers served with appropriate data and hence improving productivity of the admin as well as the user.

7.6.4.2 Disadvantages

Frequent updating of data is required, for the system to be useful.

If data have not been updated, user may get wrong results.

7.6.4.3 Applications

Applications related to organizations where huge number of personnel are required for handling user queries only.

Customer support

Taking orders

Product suggestions

7.7 Screenshots

The following screenshots provide some actual information of our implementation. Figure 7.8 depicts an interface, where user will actually interact in natural language with the system via a simple form to be filled during the sign-up activity and Figure 7.9 shows authentication done with email and password in sign-in activity.

7.8 Conclusion

We have proposed to develop a system which gives a precise and to-the-point answers to the queries of a user rather than making user search for answers amongst a huge pool of data. The system has an access to all topic-specific data and answers some personal info queries. As for natural language processing, we use Dialogflow, formerly known as

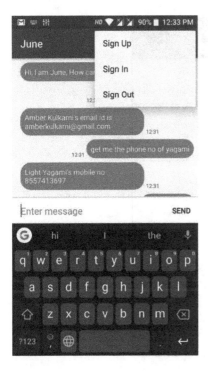 

FIGURE 7.8
Screenshot 1.

Api.ai, a service provided by Google, which helps us to set entities and its actions in the form of intents which helps us to make a chatbot which is domain specific. Thus, this domain-specific chatbot acts as an advisor to the users in a particular domain. We have incorporated an optimized context awareness to increase efficiency of the chatbot. We have included a voice-based system–user interaction that increases the usability of this system. The user asks queries in the form of voice and system gives a voice-based answer. This could thus help physically challenged people as well. We have explored other events related to the domain and found an easy way to implement it in one go. For example, consider a college conducting a survey where a notification could be given by the June that a new survey has been added to the system for a particular group of students and has to be completed. It could be then completed upon the user's approval. It will replicate the user experience as one in a personal interview. A provision can be added to provide results for non-domain-related/unanswered queries

FIGURE 7.9
Screenshot 2.

by looking into other data sources such as Google search results link, Wikipedia, and so on

References

[1] Chai and J. Lin, "The role of natural language conversational interface in online sales: a case study," International Journal of Speech Technology, vol. 4, pp. 285–295, Nov. 2001.

[2] J. Chai, V. Horvath, N. Nicolov, N. Stys, K. Zadrozny and W. Melville, "Natural Language Assistant: a dialogue system for online product recommendation," AI Magazine; ProQuest Science Journals Summer vol. 23, no. 2, pp. 63–75, 2002.

[3] G.M. D'silva, S. Thakare, S. More and J. Kuriakose, "Real World Smart Chatbot for Customer Care using a Software as a Service (SaaS) architecture," International conference on I-SMAC (IoT in Social, Mobile, Analytics and Cloud) (I-SMAC 2017)

[4] S. Ghose and J.J. Barua, "Toward the implementation of a topic specific dialogue based natural language chatbot as an undergraduate advisor",

Department of Computer Science and Engineering, University of Information Technology and Sciences Baridhara, Dhaka, Bangladesh, IEEE 2013.

[5] P. Salve, "Vishruta Patil, VyankateshGaikwad, Prof. Girish Wadhwa, "College Enquiry Chatbot", Department of Information Technology Vidyalankar Institute of Technology, Wadala (E), Mumbai, India, International Journal on Recent and Innovation Trends in Computing and Communication, vol. 5, no. 3.

[6] M. Niranjan, M.S. Saipreethy and T. Gireesh Kumar, "An intelligent question answering conversational agent using naïve Bayesian classifier", Department of Computer Science and Engineering, Amrita Vishwa Vidyapeetham, Ettimadai

[7] N.T. Thomas, "An E-business Chatbot using AIML and LSA," Amrita Vishwa Vidyapeetham University, Kollam, Kerala, India.

[8] M. Naveen Kumar, P.C LingaChandar, A. Venkatesh Prasad, K. Sumangali, "Android based educational chatbot for visually impaired people", School of Information Technology and Engineering VIT University, Tamil Nadu, India.

[9] H. Al-Zubaide and A.A. Issa, "OntBot: Ontology based ChatBot", 2011 Fourth International Symposium on Innovation in Information & Communication Technology.

Section III

Machine Learning in IoT

8

Implementation of Machine Learning in the Education Sector

Analyzing the Causes behind Average Student Grades

Prayag Tiwari, Jia Qian, and Qiuchi Li
Department of Information Engineering, University of Padua, Italy

CONTENTS

8.1 Introduction

Currently, the IT industry has been growing rapidly, and in this journey, machine learning (ML) has been playing a vital role. ML has become so popular nowadays that almost all IT industries are using it to retrieve hidden information from the massive amounts of data. Thus, it is necessary to turn these insights into applicable models, applied onto various areas, such as agriculture, transportation, banking, medical, and so on. Chatterjee et al. (2018); Dey, Pal, and Das (2012); El-Sayed et al. (2018); Fong et al. (n.d.); Hore et al. (2017); Kamal et al. (2018); Saba et al. (2016); Singh et al. (2017); Tiwari (n.d.); Tiwari, Dao, and Nguyen (2017); Tiwari, Kumar, and Kalitin (2017); Virmani et al. (2016) are some studies on this aspect.

ML can also be applied in the educational sector, as discussed by Han, Kamber, and Pei (2000). Precisely, it may predict students' categorization problem using classification methods (Brijesh Kumar, and Sourabh, 2011), such as J-48, multi-support vector machine (multi-SVM), artificial neural network (ANN), and so on. Clustering techniques can be used to study the association between students, such as fuzzy C-means (FCM), fuzzy k-means (FKM), and so on. A rich source of hidden knowledge can be discovered from this educational data, which may be beneficial for improving students' performances in the near future. The principal objective of advanced educational establishments is to provide qualified instructions to the education sectors and to enhance the nature of administrative choices. We believe that an effective approach is to collect education-related data and use it to retrieve hidden information so as to enhance the quality of education as well as provide constructive suggestions to students and teachers. ML can be utilized to provide and offer useful proposals to scholarly organizers in advanced educational establishments to upgrade their primary leadership skills and to enhance schooling performances (Tair and El-Halees, 2012).

The motivation of this paper can be summarized from several aspects:

(1) to determine the factors that influence students' performance in terms of grades;

(2) to inspect several hypotheses regarding the factors influencing students' performances; and

(3) to predict students' grades via different algorithms (regression tree (RT), linear regression (LR), and random forest (RF)).

The organization of this publication is as follows. Section 1 describes the background and points out the significance of this research, section 2 describes literature survey, section 3 describes the materials and proposed methodologies, section 4 discusses the simulated results, and the last section concludes the work and provides scope for the future.

8.2 Literature Survey

Several articles have already demonstrated that young males consume more amount of alcohol than young females. There is a significant number of high school students who consume alcohol more often weekly, leading to several problems such as becoming intoxicated, becoming increasingly frustrated and angry, having problems in social behaviors, and so on (J. V. Rachal et al., 1980). National studies on liquor use among youths (Rachal et al., 1975, 1980) and grown-ups in the overall public (e.g., Cahalan et al., 1969; Clark and Midanik, 1982) have demonstrated that drinking is more prominent in the northeast than in many other districts of the United States (Cahalan, Cisin, and Crossley, 1969; Milgram, 1993; J. Rachal et al., 1975).

A study was conducted in New York State to find out the pattern of alcohol consumption among 7th–12th-grade students. Of the total of 27,335 students in the study, 71% were found to consume an average amount of alcohol, while 13% consumed excessive amounts of alcohol, i.e., five to six drinks every week. Hispanic, Oriental, and Western Indian students were found to consume less alcohol compared to American Indian students. The rate of alcohol consumption is much higher among students than what was found in a national survey of adolescents. As obtained from the analysis of the survey, students consuming more alcohol were more likely to miss school more often, have more friends who consume alcohol, and get poor grades in school (Barnes and Welte, 1986).

Romero and Ventura Romero and Ventura (2007) conducted a study of educational data mining for the period around 1995–2005. They reasoned that education mining is a promising zone of research and it has particular characteristics that are not seen in many other areas. Thus, there is a substantial need for ML to improve the education sector.

Baradwaj, Pal Brijesh Kumar, and Sourabh (2011) implemented a classification-based data-mining strategy to assess students' potential in which they utilized the decision-tree technique for classification purposes. The objective of their investigation is to retrieve information that portrays students' potential in the end semester test. They utilized students' information from the student database, including attendance, seminar, class test, and assignment marks. This examination enables a prediction of the dropout rate and an identification of students who require extra consideration, allowing the educator to take suitable measures.

Schools are an essential part of our society to educate students. The infrastructure of a school is a very crucial indicator of the quality of the school. It consists of proper arrangement of classrooms, libraries, toilets, and so on. Schools having better results attract other students to get enrolled into that school. Jariwala, Desai, and Zaveri (2016) analyzed school data based on the infrastructure, employees, and results and found the pattern behind the raw dataset on the proper infrastructure needed by a school, which remains one of the most prominent things in a school.

P. Tiwari et al. (2017) utilized ML or deep learning methods such as Naive Bayes (NB), support vector machine (SVM), and maximum entropy (ME) on the Rotten Tomatoes movie dataset using the n-gram approach. It shows that the precision of a classifier is lessened by expanding the estimation of "n" in n-gram, i.e., it is noticed that the outcome is optimal in the case of unigram, bigram, and trigram, yet the exactness diminishes when identified for four-gram and further on.

There are expanding research interests in utilizing data mining in the education sector. This recently developing field of educational data mining mainly aims at creating strategies that concentrate on learning management systems. Our primary motivation behind this work is to introduce this to all users who are interested in this area. In this work, we have provided the full procedure of retrieving e-learning datasets and how to implement ML algorithms, i.e., clustering, association rule mining, classification, visualization, and so on (Romero, Ventura, and García, 2008).

8.3 Materials and Proposed Methodologies

This study primarily attempts to build and train a predictor that can forecast mathematics and Spanish grades by retrieving hidden knowledge from the historical data. The objective can be interpreted as a supervised regression problem; several ML methods have been widely utilized such as RT (Breiman, 2017), LR (Kanyongo, Certo, and Launcelot, 2006), and RF (Ho, 1995).

8.3.1 Linear Regression (LR)

LR aims to model the connection between variables by fitting a linear equation to the observed dataset. The set of features is combined by different weights, which imply the significance of the corresponding features. This makes it easy to understand and interpret, and this is the main reason for the popularity of LR. An appropriate preprocessing step is required before training the linear model, such as filling in the missing data, rescaling the variables, and so on. Although some may seem trivial, they end up influencing the performance. Furthermore, LR has numerous variations to mitigate the overfitting issue, e.g., least absolute shrinkage and selection operator (LASSO; Xu, Caramanis, and Mannor, 2009). We use it as linear regressors. Let us assume we have the dataset $S = ((x_1, y_1), (x_2, y_2), \dots , (x_n, y_n))$, where x_i is the p dimension vector if the number of features is p, indicated as $x_i = (x_{i1}, x_{i2}, \dots , x_{ip})$. w_i is another vector that represents the weights, $w_i = (w_{i1}, w_{i2}, \dots , w_{ip})$. The linear function is defined as f, y' is the predicted value in response to x_i.

$$y'_i = f(x_i)$$

$$f(x_i) = \sum_{h=1}^{p} w_{ih} * x_i + b$$

where f(.) is selected from a set of linear models F by minimizing the errors and satisfying the constraint, and b is the intercept (the estimation of y when x=0).

$$f^*(x) = \underset{f \in F}{\arg\min} \sum_{i=1}^{n} (y_i' - y_i)^2$$

subject to

$$\sum_{h=1}^{p} |w_{ih}| \leq t$$

where t is a parameter that is tunable; when t is large, the constraint is loose and when t decreases, the constraint becomes restricted.

8.3.2 Regression Tree (RT)

RT can be considered as a variation of decision tree (DT), which predicts the real value on leaf nodes. The internal nodes are represented by variables (features), and the break points split into branches. The break point could be a real value if the variable is numerical, whereas it could be a binary number if the value is nominal. In general, RT is built from top to bottom with the root sitting on the top, and it gradually keeps dividing until the preset condition is met, and the leaf is finally generated seated on the bottom. The same features could repeatedly be used with different thresholds to split the branches. It permits input factors to be a blend of categorical and continuous variables, bypassing the need for categorical variable encoding.

An RT works through a procedure known as binary algorithmic allotment; it is a repetitive procedure that divides the information into parcels or branches, and subsequently keeps half of each section into smaller segments as a result of which it moves up each branch. The pseudo-code has been summarized next.

8.3.3 Random Forest (RF)

RF is a method for both classification and regression tasks, often referred to as CART (classification and RT). In this paper, we will introduce the regression version. It is a collection of trees, and every tree is trained by a subset of data

---

Algorithm 1: Pseudo-code of RT

- Step 1: Fit a regression function with order two to root node minimizingab-solute deviation, recorded as ERROR^{root};
- Step 2: Initially, we consider the root node as the current node, and ERROR^{current} = ERROR^{root};
- Step 3: For every current root and for each input variable, we solve theoptimization of the regression problem. The deviation is noted as ERROR^{split};

- Step 4: Identify the best split by minimizing the loss function, $ERROR^{split} = minmERROR^{split}$;
- Step 5: If $ERROR^{current} - ERROR^{split}$ does not meet the condition, thecurrent node is split; otherwise, the current node is terminated as the leaf node;
- Step 6: For each child node, repeat steps 3–5.

---

samples and variables that are randomly chosen, instead of the whole dataset as the RT does. These trees are fully grown (low bias, high variance), and they are relatively uncorrelated by the random generation fashion. RF may alleviate the overfitting problem to some extent by reducing the variance (average the trees).

8.3.4 Dataset Description

This dataset is taken from the educational survey of Spanish and mathematics subjects student from secondary school. This dataset consists of several interesting information about students. It can be useful to analyze and investigate the performance of students. There are two datasets comprising mathematics and Spanish grades. First, we merged the students who are repeatedly present in both datasets. The average grade of mathematics and Spanish is computed as the value to predict. The dataset contains 84 students, and every record has 33 variables. We take 70% of the dataset as the training data and 30% as the test part. We have demonstrated some of them in Table 8.1, such as sex of the student (male or female), age of the student (15–24), mother's education (0–5 in which 0 indicates no education and 5 indicate higher education), father education (0 to 5 in which 0 indicate no education and 5 indicates higher education), guardian of the student (father, mother, or other), job of father (teacher, health, civil, home, or other), study-time of the student (1–10 hours), extra support for education from school (yes or no), educational support from family (yes or no), extracurricular activities (yes or no), student wants higher education (yes or no), accessibility of Internet at home (yes or no), relationship of family (1–5, from bad to best), going out with friends (1–5, where 1 is very less and 5 is very often). The dataset contains two tables of mathematics and average Spanish score, which are merged into one table for further analysis.

8.4 Results

In this section, we conduct several experiments, show the results, and give the corresponding explanation. First, we plot the grades of mathematics as the x-axis and the grade of Spanish as the y-axis.

TABLE 8.1

Table of features

Feature	Type	Range
Gender	categorical	Male, Female
Age of Student	numerical	15–24
Mother/Mother's Education	numerical	0–5
Guardian of Student	categorical	father, mother, or other
Job of Father/Mother	categorical	teacher, doctor, civil, home or other
Study Time	numerical	1–10 hours
Extra support from School	categorical	yes or no
Educational Support from Family	categorical	yes or no
Extracurricular Activities	categorical	yes or no
Desire for Higher Education	categorical	yes or no
Accessibility of Internet	categorical	yes or no
Relationship with Family	categorical	1–5
Go Out with Friends	categorical	1–5
...	...	...
To predict Feature	**Type**	**Range**
Average Grade (of Math and)	numerical	0–20

The above scatterplots (Figure 8.1) have a couple of suggestions. To start with, among the 84 students, nobody consumed high or large amounts of liquor routinely. Second, those with high scores consumed low levels of alcohol on weekdays. Third, mathematics and Spanish evaluations appear to correspond exceedingly with each other. When we relapsed Spanish assessment on mathematics grades, the balanced R-squared is 0.56. This implies that the connection coefficient among mathematics and Spanish evaluations is around 0.73 and that around 56% of the variety in Spanish evaluations can be clarified by the variety of mathematics grades. In our view, this means we can simply join the two tables together without stressing considerably over the topic, regular evaluations in math or Spanish reflect typical student scenario.

The normal middle grade is outwardly higher among those students who had low levels of daily liquor utilization. Notwithstanding, the middle grades of students with medium, higher, and very high amounts of daily liquor consumption do not appear to be altogether different. As our first cut at the consistency of standard evaluations utilizing every other variable, execute multiple LR then construct an RT of normal levels on every single other variable. The variable "failure" is firmly identified with the objective variable grades. Therefore, grades and past failure depict

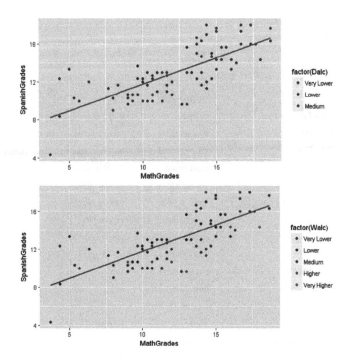

FIGURE 8.1
Scatterplot between mathematics grades and Spanish grades

a similar general student scenario, and thus we removed the variable "Failure" from our given dataset. Balanced R-squared in the above regression is just 0.18, which is very low. It suggests that exclusive 18% of the alteration in the average grades is clarified by the alteration in everything else. The factors that have factually noteworthy effects on the average grades of students considered for review are schools, study time, higher, and paid.

As per the RT analysis, the variable that is by all accounts vital is "higher," which demonstrates whether students need to seek advanced education. A greater number of the reviewed students might want to seek advanced education, and their average grade score (11.4/20) is fundamentally higher than the average grade of students who do not (8.47/20). RT analysis shows that mother's education is another vital variable. Students whose moms had any kind of secondary education had a fundamentally higher score (12.2) than students whose moms did not, having average scores (11.1/20). Next we assess the relative predictive execution of the two models. First, we carry out normalization with the mean squared error of the two models. The result showed that LR performs better than RT. This error scatterplot is shown in Figure 8.3.

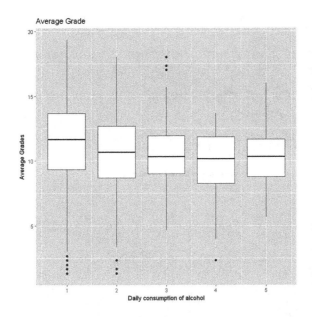

FIGURE 8.2
Scatterplot between average grades and daily consumption of alcohol

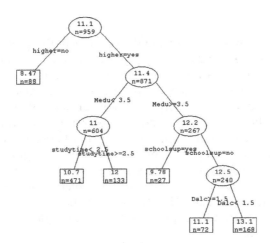

FIGURE 8.3
Error scatterplot

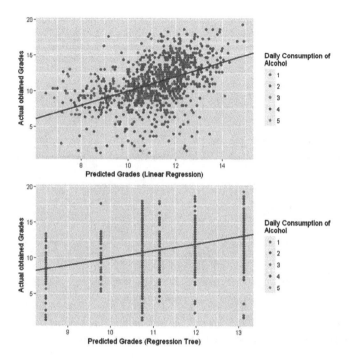

FIGURE 8.4
Scatterplot between the actual obtained grades and predicted grades

In the above charts, the horizontal axis depicts the predicted scores or grades, whereas the vertical axis depicts real grades or scores. On the off chance that the model is precise in anticipating actual scores or grades at that point anticipated grades or scores must be equivalent to real grades, and in this manner, the scatterplot should arrange along the 45 degree (blue) line. As the normalized mean squared error and error plots demonstrate, neither of the models appears to have a better than average score with regard to forecasting student average scores or grades. Unsatisfied with how RT and LR models execute, that is why RF used to go into thick. The normalized squared mean error of the implemented RF is 0.25, and it is approximately lower than RT and LR. For validation purposes, the error plot of RF is plotted and compared with the error plots of RT and LR, as shown in Figure 8.5.

Despite the fact that the RF appears to methodically underpredict the score of poor-scoring students and overpredict the scores of high-scoring students, by and large RF is by all accounts a vastly improved indicator of average scores or grades than either the RT or LR. Moreover, 10 × 5-fold cross-validation executed on a machine confirmed that the RF performed

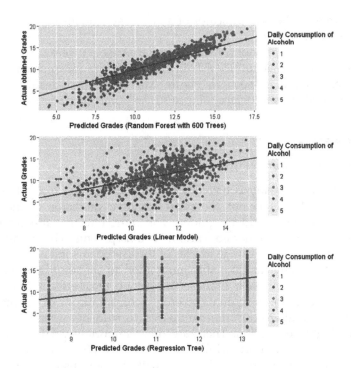

FIGURE 8.5
Scatterplot between the actual grades and predicted grades

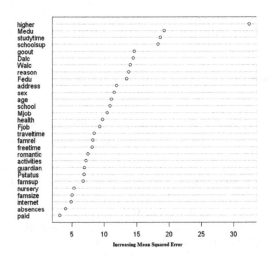

FIGURE 8.6
Scatterplot

with 600 trees is the best indicator of an average student's scores or grades compared to RT and LR.

If we remove the alcohol consumption variable on a weekday or weekend basis, there will be about 13–15% increment in the mean square error of predictions, as can be seen in Figure 8.6. Thus, these features are also important predictors in determining students' average grades.

8.5 Conclusions and Future Work

The results of our analysis have shown that consuming alcohol on weekends and weekdays is not the most effective predictor of increase or decrease in average student grades. The most effective predictor is the eagerness to pursue higher education, such as parents who motivate their children, and thus the average grades of those students increased from 9.45 to 11.4 and students without this motivation to pursue higher education showed a decrease in their grades. The second most effective predictor is the mother's education; students' average scores improve over time if their mothers have higher education compared to students whose mothers do not have higher education. There are other important predictors, such as extra educational support, which has a negative effect, as seen from the decreased average grades; going out with friends also has more often a lousy effect on average student grades; and students who consume alcohol in low amounts on a daily basis have some improvement in their averages grades compared to those who consume alcohol in high amounts. Thus, if these factors are taken into account, average grades of students can be improved. For future work, researchers can use big dataset to provide more precise results to predict the causes behind increase or decrease in average student grades.

References

[1] Balas, V. E., & Shi, F. (2017). Indian sign language recognition using optimized neural networks. In *Information technology and intelligent transportation systems* (pp. 553–563). Springer.

[2] Jariwala, D. R., Desai, H., & Zaveri, S. H. (2016). Mining educational data to analyze the performance of infrastructure facilities of gujarat state. *International Journal of Engineering And Computer Science*, 5(7).

[3] Kamal, S., Dey, N., Nimmy, S. F., Ripon, S. H., Ali, N. Y., Ashour, A. S., ... Shi, F. (2018). Evolutionary framework for coding area selection from cancer data. *Neural Computing and Applications*, 29(4), 1015–1037.

[4] Kanyongo, G. Y., Certo, J., & Launcelot, B. I. (2006). Using regression analysis to establish the relationship between home environment and reading achievement: A case of Zimbabwe. *International Education Journal*, 7(4), 632–641.

[5] Milgram, G. G. (1993). Adolescents, alcohol and aggression. *Journal of Studies on Alcohol, Supplement* (11), 53–61.

[6] Rachal, J., et al. (1975). *A national study of adolescent drinking behavior, attitudes and correlates*. Final report.

[7] Rachal, J. V., Guess, L., Hubbard, R., Maisto, S., Cavanaugh, E., Waddell, R., & Benrud, C. (1980). *The extent and nature of adolescent alcohol abuse: The 1974 and 1978 national sample surveys*. NTIS No. PB81-199267. Springfield, VA: National Technical Information Service.

[8] Romero, C., & Ventura, S. (2007). Educational data mining: A survey from 1995 to 2005. *Expert Systems with Applications*, 33(1), 135–146.

[9] Romero, C., Ventura, S., & Garcia, E. (2008). Data mining in course management systems: Moodle case study and tutorial. *Computers & Education*, 51(1), 368–384.

[10] Saba, L., Dey, N., Ashour, A. S., Samanta, S., Nath, S. S., Chakraborty, S., ... Suri, J. S. (2016). Automated stratification of liver disease in ultrasound: An online accurate feature classification paradigm. *Computer Methods and Programs in Biomedicine*, 130, 118–134.

[11] Singh, J., Prasad, M., Daraghmi, Y. A., Tiwari, P., Yadav, P., Bharill, N., ... Saxena, A. (2017). Fuzzy logic hybrid model with semantic filtering approach for pseudo relevance feedback-based query expansion. In *2017 IEEE Symposium Series on Computational Intelligence (SSCI)* (pp. 1–7).

[12] Tair, M. M. A., & El-Halees, A. M. (2012). Mining educational data to improve students' performance: A case study. *International Journal of Information*, 2(2), 140–146.

[13] Tiwari, P. (n.d.). Comparative analysis of big data.

[14] Tiwari, P., Dao, H., & Nguyen, G. N. (2017). Performance evaluation of lazy, decision tree classifier and multilayer perceptron on traffic accident analysis. *Informatica*, 41 (1), 39.

[15] Tiwari, P., Kumar, S., & Kalitin, D. (2017). Road-user specific analysis of traffic accident using data mining techniques. In *International Conference on Computational Intelligence, Communications, and Business Analytics* (pp. 398–410).

[16] Tiwari, P., Mishra, B. K., Kumar, S., & Kumar, V. (2017). Implementation of n-gram methodology for rotten tomatoes review dataset sentiment analysis. *International Journal of Knowledge Discovery in Bioinformatics (IJKDB)*, 7(1), 30–41.

[17] Virmani, J., Dey, N., Kumar, V., et al. (2016). Pca-pnn and pca-svm based cad systems for breast density classification. In *Applications of intelligent optimization in biology and medicine* (pp. 159–180). Springer.

[18] Xu, H., Caramanis, C., & Mannor, S. (2009). Robust regression and lasso. In D. Koller, C. Schuurmans, Y. Bengio, & L. Bottou (Eds.), *Advances in neural information processing systems 21* (pp. 1801–1808). Curran Associates, Inc. Retrieved from http://papers.nips.cc/paper/3596-robust-regression-and-lasso.pdf

9

Priority-Based Message-Forwarding Scheme in VANET with Intelligent Navigation

Sachin P. Godse
Ph. D Scholar, Department of Computer Engineering, Sinhgad Institutes, Smt. Kashibai Navale College of Engineering, Savitribai Phule Pune University, Pune, India

Parikshit N. Mahalle and Mohd. Shafi Pathan
Professor, Department of Computer Engineering, Sinhgad Institutes, Smt. Kashibai Navale College of Engineering, Savitribai Phule Pune University, Pune, India

CONTENTS

9.1 Introduction

Vehicular adhoc network (VANET) is the adhoc network that provides intelligent transportation services. Transportation plays a crucial role in smart city development. There are several aspects of VANET that have not been studied, and hence it is an area that provides much scope for research. In VANET, nodes communicate with each other to share either safety or non-safety information. Safety information may contain various kinds of messages like traffic jam condition, accidents met by vehicles, bad weather, road conditions, collision of vehicles because of sudden break applied by the next vehicle, and so on. Non-safety information may include messages like driver details, documents, songs, movies, and so on. In the event of a fire accident in the city, an automatic call gets placed to the fire station due to alarm systems or because of someone manually calling for a fire brigade. As the fire brigade has to reach the accident location as soon as possible, it is considered as an emergency event. This emergency event message is forwarded to all vehicles in the range of the fire brigade, so that they can give it way, creating an easy and fast passage for the fire brigade. If the road to the place of the fire accident is packed due to traffic, then the intermediate vehicles can suggest an alternate path to reach that place. The driver can also make a request for an alternate path to the server, which, on being provided, can be used by the driver to select the best path and proceed to the accident location. It is very important for a fire vehicle to reach the location of fire on time to avoid fatalities and property losses. Figure 9.1 shows a VANET scenario and the types of communication associated with the scenario [1][2].

9.2 Computational Intelligence in VANET

Transportation services play a crucial role in the development of any country. Currently, there are numerous global problems associated with transportation like traffic jams, increase in accidents, misbehaviors on road

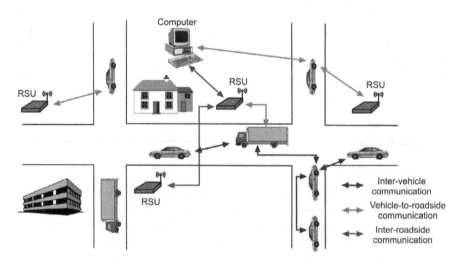

FIGURE 9.1
Vehicular network scenarios.

by a driver, no prior information about weather and road conditions, and so on. VANET addresses these problems by imparting intelligence in transportation through which vehicles communicate with each other and share important information among them, which in turn assists a driver in taking decisions at runtime. Vehicles are deployed with various environmental and necessary sensors on them, thereby enabling continuous access of information. The collected information is given to the on-board unit (OBU), which is deployed on the vehicle. The OBU processes the collected information and generates messages based on that. OBU forwards these messages to other vehicles or Road Side Unit's (RSU's). Information of city map, roads, vehicle personal information, and certificates are available with the trusted authority (TA). This information can be made available if required in the network through RSUs. Information/ messages in VANET can be categorized into two classes of services: safety messages,which is associated with safety in driving by forwarding alert messages generated during emergency situation; and non-safety messages, which are related to the comfort of owner/driver by forwarding non-safety messages like navigation, path, movie, songs, and so on. Data-related messages are provided on request by a vehicle. Intelligent navigation plays an important role in runtime vehicle driving. It provides path information to a driver, suggests the shortest path to a driver, and also an alternative path for the destination if an emergency situation arises in the current path. There is much research under way in analyzing traffic situations at runtime by observing the density of vehicles on roads. VANET plays an important role in emergency situations like accidents, natural calamities, weather conditions, and so on by communicating and processing this information, and thus has a huge scope for research in the future.

In this chapter, we address issues in communication and navigation of VANET. Communication in VANET is provided by WAVE (Wireless Access for Vehicular Environment), that is, IEEE 802.11P standard. It is an extension of IEEE 802.11. DSRC (Dedicated Short Range Communication) gives separate 75 MHz of spectrum in the 5.9 GHz band in the USA. VANET is very dynamic in nature, where nodes/vehicles move very fast, as a result of which vehicles get very less time to interact with each other. Communication efficiency can be improved by sending/forwarding only important messages by vehicles or RSUs by assigning priority to the messages. Priority-based message forwarding is the first objective of this chapter. Second, in emergency situations, the system should provide suitable, time-saving alternate paths by skipping crowded roads or roads with traffic jam. The remaining chapter is organized as follows. Section 2explains the existing schemes in message forwarding in VANET; section 3 elaborates the proposed message-forwarding scheme; section 4 proposes a navigation scheme for VANET; section 5 discusses the results; and section 6 concludes the chapter.

9.3 Existing Schemes in VANET

In [3], the author discusses the issues related to developing certain protocols that are needed for autonomous and robust broadcasting purposes. These provide each node with sufficient strategy for determining whether the inbound message has progressed further or not. It depends on the priority level and on the overall network load. The objective here is to make the best use of wireless resources when messages are sent concurrently. The author provides techniques for broadcasting, which are employed for sending messages properly. When a packet is sent, it is accepted by all the nodes in the sender's coverage region. Each receiver then decides to resend the packet based on its individual sending strategy. For example, if any critical message such as an accident alert is to be delivered, it should be done as soon as possible. However, it does not make much difference whether some weather-related data is broadcast late.

In [4], the author focuses on providing differential service to different priority safety messages in a VANET with the following characteristics. i) All vehicles move fast; thus, it can cause high bit error rates. ii) All vehicles move in the same direction, following the road topology. Therefore, the relative velocity of all nodes is comparatively low. iii) The communication safety messages have a few hundred bytes and have different priorities. To increase the probability of a successful transmission for high-priority messages, the message should be transmitted more times than its counterpart, each with a relatively low priority.

In [5], the author has designed mobility-centric approach for data dissemination in vehicular networks MDDV to address the data dissemination problem in a partitioned and highly mobile vehicular network. Messages are forwarded along a predefined geographic trajectory.Since no end-to-end connectivity is assumed, intermediate vehicles must buffer and forward messages opportunistically. As an opportunistic algorithm, MDDV addresses the question about who can transmit, when to transmit, and when to store/drop messages. Using a generic mobile computing approach, vehicles perform local operations based on their own knowledge, while their collective behavior achieves a global objective.

In [6], the author addresses a priority-based exigent data dissemination approach for time-sensitive traffic of VANET where a laxity-based priority scheduling technique with a back-off mechanism is introduced. The scheme selects the next packet to be transmitted based on a priority calculation function that takes into consideration the uniform laxity budget of the flow, the current packet delivery ratio of the flow, the desired packet delivery ratio, and the request selection precedence value.

In [7], the author provides a priority-based vehicle movement. This network gives an emergency vehicle a high priority, registered vehicle a medium priority, and unregistered vehicle a low priority. To avoid an accident, the Chord algorithm is used to control the speed of the vehicle. It measures the speed of the vehicle, and intimates it when it crosses the speed limit to avoid accidents.

In [8], the author recommends a new approach for path guidance across the Internet for the purpose of collecting useful information. This approach also provides proper guidance to new users for reaching the target. It can obtain real-time data of traffic conditions and congestion caused by unforeseen incidents besides providing proper guidance to the user. The scheme also ensures that the data provided is valid to the context without affecting the credentials of users. A third party takes care of the user's privacy. Processing delay is a parameter against which the scheme is assessed.

In [9],the authors have suggested ways of improving the security and efficiency of routing. Out of the numerous factors that have been considered, beacon frame size and counts of reception are the primary ones. Increase in the size of the beacon improves all the parameters. The author also has proposed a mathematical model along with proof of concept (POC).

In [10], the authors have provided an option of using Bluetooth Low Energy (BLE) in VANET applications. They addressed the issue by employing Wi-Fi in vehicle-to-vehicle (V2V) communication considering that 90% of the smart phones are Bluetooth enabled. The authors also elaborated on some applications of VANETs with BLE. They discuss how these applications can be transformed into new ones. The results indicate that BLE communication has minimum latency and an appreciable distance between two moving objects. However, establishment of trust between two vehicles is crucial in VANETS.

In [11], the author provides a trust-based and categorized messaging scheme. A hybrid approach was used that leverages role-based and experience-based trust. Along with the hybrid approach, data-mining concepts are also used for better decision making. In addition, the authors also provide an extensive literature review along with gap analysis.

In [12], the author elaborates on the parameters to be considered in V2V communication. The parameters are studied considering security aspects. The paper discusses the need of a strong algorithm for accepting or rejecting messages. It also explains a model detecting confidence on security infrastructure (CoS) of VANET. The paper also provides a number of packets from both compromised and non-compromised nodes for analysis. Malicious nodes can be identified by using certificate revocation mechanisms.

From the literature survey discussed above, we can conclude that some schemes lack end-to-end connectivity, and retransmission of packets is required in many situations where there exists a low-density network. In high-density networks, rapid dissemination causes a large number of collisions. Some schemes however do not allow unregistered vehicles to take advantage of the proposed mechanism or else give them less priority. It also requires the employment of the vehicle traffic flow theory to transmit a time-consuming packet. Many schemes consider low-density, high-density, and medium-density vehicular network separately. The scheme does not consider the validity of messages by checking the time of generation/arriving in the network, which may cause wastage of time by forwarding older messages. The priority scheme should work at both levels of RSU and Vehicle, which is unfortunately not addressed in any scheme.

Considering the above limitations, the scheme proposed in this chapter deals with vehicles in high-, low-, and medium-density environments. This scheme discards older, irrelevant messages to a current scenario by setting a time threshold value, which increases efficiency by avoiding wastage of time. Rapid dissemination of messages is avoided by maintaining a constant bit rate for all messages, and the sequence in which messages are to be forwarded is regularized by following a priority-based message-forwarding mechanism.

This chapter also provides experimentation results by applying the proposed scheme to emergency situations using an intelligent navigation mechanism. Here, the VANET simulator (Vsim) for simulating the proposed scheme for emergency situation navigation and path changing has been used.

9.4 Proposed Message-Forwarding Scheme in VANET

Messages in VANET are mainly classified into two types—safety messages and non-safety messages. Figure 9.2 shows the different types of messages in VANET. As shown in the diagram under safety messages, all emergency messages are listed, and under non-safety messages, all other information having lesser priority than safety messages are listed.

Figure 9.3 shows the steps for priority-based message forwarding. Step 1 categorizes the messages depending on their type. There are two main categories of messages—safety and non-safety, as shown in Figure 9.2. Depending on the nature of the incoming message, its type is set by setting the bits based on the message packet format. Step 2 sets the priority of different types of messages categorized in step 1. The priority levels are as

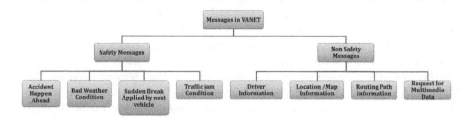

FIGURE 9.2
Classification of messages in VANET.

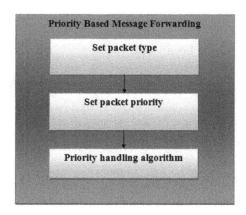

FIGURE 9.3
Steps for assigning priority to message.

follows: 0, 1,2,3,4, and so on, with 0 being the highest priority. Step 3 applies the priority handling algorithm as explained in the following section.

9.4.1 Relevance/priority-based message-forwarding scheme

Table 9.1 lists the relevance/priority assigned to messages depending on their severity for broadcasting in the network.

TABLE 9.1

Message Priority

Sr. No.	Relevance/Priority	Message Type
1	0	Emergency vehicle(e.g., ambulance)
2	1	Road accident
3	2	Sudden break applied
4	3	Traffic jam
5	4	Bad weather
6	5	Driver information
7	6	Location or map information
8	7	Routing path information
9	8	Request for multimedia data

Parameter

1. Event Sense
2. Destination Vehicle Location x, y
3. Distance
4. Vehicle Speed

Algorithm [13]

1. **Generate event (message) by source node.**

   a) Sense event.

   b) Assign priority to event.

   c) **If** (destination node is in the range of source)

      i. Source node broadcast message.

   d) **Else** (message forwarded to destination through RSU)

      i. RSU verifies message.

      ii. RSU stores packet for time t.

      iii. RSU applies the scheduling algorithm.

2. **Forward message by receiver node**

   i. Check priority of received messages.

   ii. Forward messages in sequence of their priority values, 0, 1, 2, 3....8.

   iii. Sort messages in ascending order of distances of nodes (i.e. less distance first).

   iv. Forward messages one by one.

3. **Schedule vehicle and RSU**

   a) **If** (message waiting time < threshold time "t")where message waiting time = current time − sense time.
   Forward message

   b) **Else**

   Remove packet from node database.

4. **Forwarding—Vehicle**

   a) Forward message by the receiver node algorithm.

5. **Validate message**

   a) Source node validation by the receiver node.

   b) Message validation.

   c) **If** (valid node and valid message)
   Forward message.

   d) **Else**

   Drop message.

9.5 Proposed Intelligent Navigation Scheme

Figure 9.4 shows the proposed architecture for scheme implementation. The first module is the vehicle module, in which vehicles get authenticated. If any event (e.g., accident, traffic jam, road condition) occurs within the selected route of the vehicle, then it detects that event and broadcasts the message to other vehicles and to the RSU; other vehicles receive the message from the vehicle that initiated the alarm and makes changes in the route if required. At the server side, the message (event packet) is received and checked against the priority list in the server. The priority list of messages is created on the server side. According to the priority of the messages, messages are broadcast to all the vehicles. Other vehicles then detect that event and broadcast the same event to all nearby vehicles and RSU.

If a vehicle requests for a path, then the server checks the paths with respect to the source and destination in a database. An alternate path is selected by skipping the accident-prone paths, which is then sent to the vehicle. Vehicle security is maintained by first authenticating it, followed by which vehicle information is saved in the database. The authenticated vehicle shares a public key with the certification authority and generates its own private key. Similarly, the certification authority shares the public key. The vehicles communicate with other vehicles via a wireless medium like Wi-Fi. Each vehicle that participates in the VANET is deployed with an OBU. All interactions between the vehicles and the RSUs happen through the OBU. The OBU processes the messages received from other vehicles and the data from various sensors on the vehicle, thereby playing an important role in path discovery and event management. Vehicle authentication is carried out using

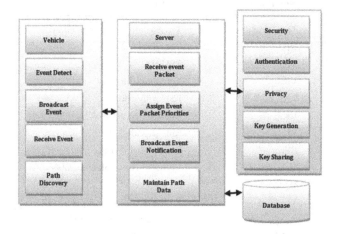

FIGURE 9.4
System architecture for navigation.

the RSA/ECC algorithm. Key generation and key sharing/exchange happen at the server end. Keys are sent to vehicles. The server provides path information, path discovery, and event priority assignments. Rerouting can be initiated at run-time in case of emergency events.

Algorithms

The algorithm used has three different phases—the initializing phase, the authentication phase, and the navigation phase. Table 9.2 describes the different notations used in the algorithm.

A) Initializing Phase: In this phase, the vehicle and the RSU register with the TA. After registration, the TA shares some identity parameters with the vehicle and the RSU, which is included in the certificate.

1) TA generates self-ID and public key TK_{pub}.
2) Set key size (K_{size}) and generation point (GP).
2) Vehicle registers with TA by submitting information of the user and vehicle (vehicle unique chassis number, vehicle type, user details).
   $V_i \rightarrow$ (Unique ID, VID, VK_{Pub}) i= 0 to n (n = number of vehicles).
3) RSU registers with the TA by providing its unique ID and location.
   $R_i \rightarrow$ (Unique ID, LR_i)

B) Vehicle Authentication: After the registration phase, the TA assigns a vehicle ID and a secret key to each vehicle.

TABLE 9.2

Notation Used in Algorithms

Notation Used	Meaning
Tk_{Pub}	Trusted authority public key
VID	Vehicle ID
RID	RSU ID
VK_{Pub}	Vehicle public key
RK_{Pub}	RSU public key
VK_{Pri}	Vehicle private key
RK_{pri}	RSU private key
LR_i	Location of ith RSU
V_{LPN}	Vehicle license plate number
V_k	K^{th} vehicle
$V_{kPosition}$	K^{th} vehicle's current position
V_{kDest}	Destination of Kth vehicle
EnMsg	Encrypt Message
DeMsg	Decrypt message

1) TA $\rightarrow$ (VID,VK$_{pri}$)
2) TA assigns a license plate number to each vehicle.
   TA $\rightarrow$ V$_{LPN}$
3) TA generates and assigns certificates to each vehicle
   CV$_i$ = <VID, V$_{LPN}$, VK$_{pub}$>

C) RSU Authentication: After the registration phase, TA assigns an RSU ID and a secret key to each RSU.

1) TA $\rightarrow$ (RID,RK$_{pri}$)
2) TA generates and assigns a certificate to RSU
   CR$_i$= <RID, RK$_{pri}$, LR$_i$>

D) Navigation Schemes

1) $\{V_1,V_2,.......,V_n\}$ $\rightarrow$ ReqN (RID,V$_{kPosition}$,V$_{kDest}$);
2) EnMsg = (Encrypt(RSUverification(V$_k$(msg)),VK$_{pri}$));
3) RSU $\rightarrow$ send (EnMsg);
4) TA decrypts the message DeMsg(msg,VK$_{pub}$);
5) Verify data using VK$_{pub}$;
6) Check (V$_{kDest}$,V$_{kPosition}$,VID);
7) Apply rerouting for navigation.

9.6 Navigation with Respective Conditions A and B

A) Vehicle already arrived at destination.
B) Vehicle is caught in traffic due to some emergency situation, traffic jam, and so on.

1. If (A)
2. Stop (end the navigation process)
3. Else
4. If (B)
5. Check (alternate lane)
6. Else
7. Check (alternate path)
8. Generate event/forward emergency message received
9. Broadcast event
10. If (request for path)
11. TA forwards message to RSU
12. RSU checks path

9.7 Results and Discussion

VANET Scenario using Vsim: Vsim is a java-based simulator for testing, analyzing, and implementing different protocols in VANET. We can add, create, and modify a scenario in the simulator. Vsim provides different packages for creating a VANET environment. We can load a map of different cities and load different scenarios for the same map. Figure 9.5 shows uploading of a map (NewYork_noTS.xml) in Vsim. We can create our own map depending on our requirements and upload the same onto the simulator.

Figure 9.6 shows uploading of NewYork road scenario with 2500 slow and 2500 fast vehicles. Each vehicle has a 100m communication range. Road Side Units Communication range is500m. Vehicles in motion are shown as blue dots and RSUs as gray circles in the simulator.

Figure 9.7 shows the vehicle path before the detection of an emergency event. This shows a normal flow of traffic on the selected path. The blue line shows the flow of traffic.

Figure 9.8 shows the vehicle path when an emergency event is detected. The blue line in the figure shows a change in the direction of flow.

Figure 9.9 shows the time required for a vehicle to reach the destination using priority-based messages and without priority-based message forwarding. The time required for the vehicle is listed on the y axis and the vehicle number is listed on the x axis. Table 9.3 highlights the actual values of time required for vehicles in minutes.

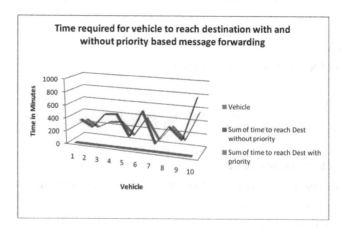

FIGURE 9.5
Times Required to Reach Destination With and Without Priority.

TABLE 9.3

Time required for vehicles to reach the end point with and without message relevance/priority.

Vehicle	Time to reach destination without relevance/priority	Time to reach destination with relevance/priority
1	313	260
2	213	155
3	435	258
4	454	280
5	125	89
6	543	387
7	56	56
8	345	295
9	165	102
10	813	546

9.8 Conclusion

The proposed scheme of priority-based message forwarding reduces the time required for message forwarding by prioritizing the emergency messages for broadcasting. It also considers the nearest receiver first for sending messages. As priority assigning takes place at both levels—the vehicle and the RSU—it is more efficient than other schemes. Efficient and effective transportation can be achieved using an event-based path finding navigation system. As the path is changed at run-time, emergency situations are handled more effectively. Because of the priority-based message-forwarding scheme, vehicles save the time spent in processing the lower-priority messages. Vehicles reach their destinations in lesser time compared to the time required to reach without the application of the proposed scheme. Here, intelligent navigation was achieved using vehicle position data, emergency event data, and traffic scenario gathered by RSUs. As the time threshold value is used to decide the freshness of a message, older messages are discarded and the time spent in processing irrelevant messages is saved. Selection of the shortest path among the available alternate paths can be the future scope of this scheme for which we propose to use a distance factor or traffic situation analyzer along the path in consideration.

References

[1] Sabih ur Rehman, et al. "Vehicular Ad-Hoc Networks (VANETs)-An Overview and Challenges". Journal of Wireless Networking and Communications 3, 3 (2013): 29–38.

[2] Sachin Godse, Parikshit Mahalle, "Time-Efficient and Attack-Resistant Authentication Schemes in VANET" Proceedings of 2nd International Conference, ICICC 2017 Springer, 579-589.

[3] Wahabou Abdou, Benoˆıt Darties, Nader Mbarek,"Priority Levels Based Multi-hop Broadcasting Method for Vehicular Ad hoc networks". Annals of Telecommunications 70, 7–8 (August 2015): 359–368.

[4] Chakkaphong Suthaputchakun et. al. "Priority Based Inter-Vehicle Communication in Vehicular Ad-Hoc Networks using IEEE 802.11e", IEEE 2007, 2595-2599.

[5] Hao Wu and Richard Fujimoto et. al., "MDDV: A Mobility-Centric Data Dissemination Algorithm for Vehicular Networks", VANET'04, October 1, 2004, 47-56.

[6] Chinmoy Ghorai et. al., "A Novel Priority Based Exigent Data Diffusion Approach for Urban VANets", ICDCN 17 January 4–7, 2017

[7] A. Betsy Felicia et. al. "Accident Avoidance and Privacy Preserving Navigation System in Vehicular Network", International Journal of Engineering Science and Computing, March 2016.

[8] Chim, Tat Wing, et al. "VSPN: VANET-based secure and privacy-preserving navigation" IEEE Transactions on Computers 63, 2 (2014): 510-524.

[9] Carpenter, Scott E. "Balancing Safety and Routing Efficiency with VANET Beaconing Messages." National Highway Traffic Safety Administration, Fatality Analysis Reporting System (FARS) Encyclopedia, October 15, 2013.

[10] Frank, Raphaël, et al. "Bluetooth low energy: An alternative technology for VANET applications". Wireless on-Demand Network Systems and Services (WONS), 2014 11th Annual Conference On. IEEE, 2014

[11] Monir, Merrihan, Ayman Abdel-Hamid, and Mohammed Abd El Aziz. "A Categorized Trust-Based Message Reporting Scheme for VANETs". Advances in Security of Information and Communication Networks. Springer Berlin; Heidelberg, 2013. 65-83.

[12] Rao, Ashwin, et al. "Secure V2V communication with certificate revocations". 2007 Mobile Networking for Vehicular Environments. IEEE, 2007.

[13] Sachin P. Godse, Parikshit N. MAhalle et al. "Rising Issues in VANET Communication and Security: A State of Art Survey". (IJACSA) International Journal of Advanced Computer Science and Applications, 8, 9 (2017), 245–252.

Section IV

Machine Learning in Security

10

A Comparative Analysis and Discussion of Email Spam Classification Methods Using Machine Learning Techniques

Aakash Atul Alurkar, Sourabh Bharat Ranade, Shreeya Vijay Joshi, and Siddhesh Sanjay Ranade
Department of Computer Engineering, Smt. Kashibai Navale College of Engineering, Pune, India

Gitanjali R. Shinde
Centre for Communication, Media and Information Technologies, Aalborg University, Copenhagen, Denmark

Piyush A. Sonewar, and Parikshit N. Mahalle
Department of Computer Engineering, Smt. Kashibai Navale College of Engineering, Pune, India

CONTENTS

10.1 Introduction

Email communication is by far the most easy, ubiquitous, and convenient medium ever used by humankind for communicating with people. It is a useful, fast, and efficient way of sending various types of data through web protocols anywhere in the world. Data sent in email includes text messages, files such as .jpeg and .png images, audio and video files, pdf and word documents, and many other such attachments. This method of transmitting information, which has been in use since the 90s, makes the geography of the sender and receiver irrelevant. It overcomes the shortcomings of other traditional communication methods such as mail by sending messages instantaneously. It is often a better alternative to fax as paper is not involved. The swift learning curve also helps. Owing to its versatility and ease of use, in the last thirty years email has pervaded the academic and corporate world. Everyday an estimated 205 billion emails are sent [15], which goes to show the extent to which emailing has come to dominate our lives. It can be argued that email, through instant communication, has played a part in the rapid development of the global economy, especially in the early 2000s, by assisting the communication among people in different countries. This means that—like any technology that permeates our lives this deeply—it is liable to be misused. This is done with both malicious and non-malicious intentions, but it is done in innumerable ways. As most people are aware, it is not possible for 200 billion emails to be manually sent. Obviously, most of these have been written by scripts and/or auto-mated bots and are used for advertising. Such types of spam emails are used for ransomware, advertising, fake purchase receipts, phishing, and increasing traffic. It is clear that the possibilities of misuse of these technologies are vast, and every year individuals and companies lose millions of dollars because of such attacks. Apart from the damage due to phishing and malware, spam emails cause a loss of around 20 million dollars every year, partly due to employees wasting company time in deleting and reading them and partly due to information being stolen from emails [1].

A survey on email frequency and spam was carried out by a US-based research firm, where it was found that around 18.5% of the total emails sent every day are of minimum importance to the receiver and 22.8% of the emails are sent unnecessarily. Spam emails cause additional load on the servers provided by email clients, and hence the minimum default storage space provided to every user had to be upgraded, resulting in incurring additional costs to either the provider or the company. With useful emails being in the same list as a spam email, it is more probable for a client to accidentally delete or ignore important emails, unconsciously doing that because they may resemble a typical spam email (for example, if the email ends with an exclamation mark, people are less likely to open it).

This is why effective spam classification is important. Although it may seem trivial and unimportant, an effective spam classifier helps the user —or any organization, for that matter—in three basic domains: security, storage space, and productivity. A simple use case discussed in the motivation section below describes how easy it is for people to fall for basic phishing attacks. Phishing refers to the attempt to obtain sensitive information, often for malicious reasons, by disguising oneself as a trustworthy entity in an electronic communication. [2] It is difficult to differentiate an authentic email from a fake one, and even experts have sometimes fumbled in these areas when the hackers were extremely skilled at their job. An efficient classifier helps in three ways: first, it prevents such emails from being exposed in the first place, thereby increasing basic security; and second, it helps in productivity. As already discussed, employees end up wasting valuable company time sorting through their inbox searching for important messages. A good classifier eliminates this need completely.

The third domain—storage space—by which a spam classifier helps us is worth mentioning because it is not immediately obvious as to why and how. It might seem that a few spam emails on a single email account would hardly take up a few kilobytes, which might be right. However, considering the enormous number of email users globally, with each email client (such as Gmail, Yahoo mail, Hotmail, and so on) servicing millions of individuals, the space taken up by spam emails now compounds exponentially, requiring companies to allocate unnecessary, unproductive larger server spaces for each employee. At first glance, this might not seem like a problem from an individual's perspective; however, given the large scale of the email networks, if a spam filter performs even as much as 1% better than another, it means thousands of emails being classified as spam and thus getting eventually deleted. Given the importance of a good classifier, it is necessary to determine the approach to develop such a filter. One that uses machine learning algorithms performs better because it iteratively improves performance with each new dataset. However, for one that uses a basic neural network model, with each epoch, by modifying the weights and iteratively processing the dataset through the same network, the results get increasingly closer to the ideal classification required.

We thus consider different algorithmic approaches for this problem. Python libraries have various predefined machine learning algorithms that can be used to classify large datasets, simply by importing the respective libraries such as numpy, TensorFlow, and pandas [3]. The next few sections of this chapter discuss the importance of machine learning in spam classification along with the various approaches used and then delineate the working of the algorithms proposed, comparing their pros and cons over various datasets. A basic workflow for a sample proposed model has also been provided. Readers can thus decide on the suitable algorithm for themselves, considering all the parameters of their unique requirement or situation.

10.2 Approaches of Email Spam Classification

Emails as a form of communication became popular when Internet service providers (ISPs) such as AOL, Prodigy, and CompuServe provided each user with an email account by default in the 1990s. This led to a surge in email usage as a primary form of communication, which complemented the increase in the number of personal computers at both offices and residences. The increasing popularity of email also saw an increase in the number of problems regarding such a method of communication. Misleading mails were on the rise, and telemarketers started flooding people's inboxes with irrelevant emails. Despite this, the popularity of this form of communication had become so huge that it did not deter people from using this service. With the increase in the number of people using emails, the frequency of emails increased. Some of the methods used to classify incoming messages have been described subsequently.

10.2.1 Manual Sorting Method

In the late 1980s, email sorting was performed manually in offices, which consisted of a person reading each email and forwarding it to the relevant department. This use case is not reliable and scalable currently as the volume of email has increased exponentially over the years. The manual sorting method also did not consider the domain of the sender as a possible feature. This posed a security threat, as the body of the email was not enough to provide a secure and convenient email experience.

10.2.2 Simple Keyword Classification

A solution to the above-discussed problem was implemented by noticing a pattern in incoming spam and marketing emails. The incoming email was classified and forwarded to the relevant people and departments by parsing the email body and searching for keywords that were previously embedded. This method was undoubtedly better than the manual sorting method, but posed a threat to scalability and understanding the different contexts. This method did not learn from previous incoming mails to consider the relevance of the present email. It only acted from the predefined bag of words, which was limited and not updated frequently. There were many disadvantages to this approach: the accuracy percentage of the classifier did not improve even after processing thousands or potentially millions of emails; the system could not be scaled up for enterprises, as it took a significant amount of time to parse through each email; and most importantly, the recurring inaccuracies could not be minimized by the absence of a feedback loop.

10.2.3 Email Aggregation Using Data Science Approaches

With the rise in the adoption of data analysis and machine learning fueled by Google open sourcing, most email classifiers started using some variation of this method. Its primary machine learning library, TensorFlow, supports other machine learning libraries, such as numpy, sci-kit learn, pandas, and so on. It has been developed in collaboration with DeepMind, which improves the decision-making significantly. The bag of words, which earlier simply parsed through entire emails, now could be embedded in a neural network layer, with each node being a feature about the email structure. This improved the accuracy percentage of the email spam classifier and was easier to develop than outsource. Using neural networks changed the percentage of spam emails arriving in a user's inbox. No spammer could game the system as all the previous inputs were considered as a "learning factor" in deciding whether the next email was spam or not. These neural networks include different features of the email body, such as header, footer, Cc/Bcc, the frequency of emails being sent, and primarily the body of the email. This resulted in a higher accuracy than with previous methods, with a more maintainable and scalable system for email management.

10.3 Importance of Machine Learning

Machine learning is currently one of the most-discussed fields in computer science. Coupled with artificial intelligence, it enables a machine or software to make intuitive decisions based on the training and testing phases in its development. [4] Besides classification, it is employed in tasks like robotics, artificial intelligence, building smart systems, gaming, image processing, and so on. It comprises two basic areas: supervised and unsupervised.

Supervised machine learning is used when the developer knows the ideal output required. Here, the training data, which is already labeled, is fed to the model, and through a certain number of iterations of testing, the model learns to identify the same labels for new unseen cases of data. The learning algorithm does this by generalizing from the training data to unseen situations in a reasonable way. Classification and regression are the two most common instances of supervised learning. Unsupervised learning is used when there is no inherent ideal output to be obtained. Here, the training data is not labeled, and the developer needs to build a model that can find certain hidden "structures" within the data and extract relevant knowledge from it. The accuracy of the structure that is output by the relevant algorithm cannot be evaluated here since there is no expected output. The most common instance of unsupervised learning is clustering.

When an email needs to be classified as spam or ham, the desired output should be definite and tangible. Every incoming email will be classified only as one of those two classes. Moreover, the training datasets that are used (such as Enron and Ling) are labeled, meaning for every email in these datasets, there is a corresponding label that states whether it is a spam or ham. All these evidence indicate that a supervised model would be the best fit. A model producing results that are the closest to the ideal must be developed for this. One huge advantage of machine learning is the sheer availability of resources online. Google's TensorFlow library, designed especially for machine learning and neural networks, is completely free along with its visualizing component, TensorBoard. Similarly, python provides libraries such as pandas, sci-kit-learn, and matplotlib that help us not only implement complex algorithms easily but also graph results for better understanding. Thus, implementation is made less complex and more available to developers.

10.4 Motivation

To reiterate the need for a sufficient classifier, it is useful to imagine a world where spam classifiers do not exist. Around two hundred billion messages continue to get sent daily, and sorting out the important ones necessitates users clicking each one and deleting them by themselves. Just imagining such a situation can be stressful. This is because a spam classifier has quietly become one of those features that we have collectively begun relying on without even noticing its presence. Another exercise would be to open any Inbox and go to the spam section. The hundreds of spam emails, advertisements, offers, and so on in that section makes one realize how much clutter has been hidden from the user's eyes. Also consider, for example, a situation where a layperson receives an email, asking for update of the user's bank account details via a provided link, which does not appear suspicious to the user because it appears to have been sent by the user's bank and looks authentic. The user then visits the link and updates all details, unaware that she/he has been the most recent victim of a phishing attack. Upon realizing this mistake, the user contacts the required authorities (such as the cyber security cell of the state) who may or may not succeed in remedying the situation and identifying the perpetrators. Now imagine if, in the same situation, the user's email client employed a more accurate spam filter. The email from the "bank" would be identified as inauthentic, based on features such as the authentication of the email id, the number of HTML tags used, spam keywords, and so on. Basically, it would go straight to spam and get deleted a few days later. The user would never find out she/he received this email, and hence there would be no damage.

The above use case illustrates just one of the ways in which spam filtering improves our lives, by simply hiding otherwise dangerous emails from our eyes. Obviously not everyone would fall for scams like these, but millions of dollars of damage are incurred each year due to other less-obvious schemes. A filter helps in these situations by providing a basic level of security. One that uses machine learning will also be able to identify such messages with greater accuracy.

10.5 Literature Survey

In this section, we will discuss previous research. Many of the proposed approaches use simple neural networks. However, no advanced deep techniques have been put forward in previous research papers. There is also no standard on how to evaluate email ranking systems, and it is observed that the results vary widely depending on the dataset. Hence, we focus on how the tasks have been approached and the algorithms/techniques employed by each of the papers. These methods tend to make use of varying levels of custom features, including ones related to dates/time, salutation, header fields, HTML tags, and the presence of questions in the text. Some studies tried to prioritize emails by categorizing the ones requiring an action. Others have stepped back to focus on the core functionalities that people use with emails and have provided innovative ways to solve email overload, such as executive assistant crowdsourcing to keep the system updated as the patterns and keywords keep on changing [5]. Classification by user folder has been approached with the use of support vector machine and wide-margin Winnow. In this paper, we intend to focus on work regarding email reply prediction strategies, as well as research dedicated to alleviating the problem of "email overload and prioritization." Because email is one of the most used communication tools in the world, Sproull and Kiesler provided a summary of much of the early works on the social and organizational aspects of email. [6] Mackay noted that people used email in highly diverse ways, and Whittaker and Sidner further studied this aspect. They found that along with basic communication, email was "overloaded" in the sense of being used for many tasks—communication, reminders, contact management, task management, and information storage. Mackay also noted that people could be divided as belonging to either of the two categories when handling their email: prioritizes or archives. [7]

Sahami et al. reported good results for filters that used naive Bayes (NB) classifiers [8]. Subsequently, many experiments showed similar outcomes, confirming Sahami's conclusion. Similar results were found by Zhang et al. related to spam classification, which incorporated machine learning algorithms. A significant change in the approach was made when the importance of both header and body for classifying mails was understood. [9]. However, more difficult problems starting cropping in spam classification with increased attacks on the filtering algorithms themselves. One out of four identified attacks by Wittel turned out to be a tokenization attack, which works based on the manipulation of the text characters, spaces, and HTML tags. These attacks work against the statistical nature of the classifiers [10]. This was eventually overcome by Boykin by utilizing interconnected networks [11]. Gray and Haahr proposed a joint spam filtering method [12]. Goodman et al. summarized other advances except machine learning in spam [13]. Although not directly used for spam classification, Raje et al. developed an algorithm for extraction of key phrases from documents using statistical and linguistic analyses. Their method focused on three areas, out of which two are relevant for spam classification on the body of the email. In the first method, the authors maintained a list of important words in the English language itself. This file was maintained as a list of words and their respective multipliers ranging from 1 to 10. The multiplier is dependent on how important the word is, and hence can be considered as the priority of the word in the language. For example, words such as "firstly," "secondly," and "therefore" are used to state new points or to conclude, as is the case for "therefore." It is obvious that phrases with such words must be given the highest consideration and so will be given a multiplier from 8 to 10, whereas other important words such as "state," "consider," or "analysis" are important but not as much as those mentioned before. The authors also proposed an idea of a dynamic list [14].

Email spam classification is the most researched issue among all the problems regarding emails. In a survey by Awed and ELseoufi, most studies on email classification are conducted to classify emails into either spam or ham. Among the 98 articles, 49 are related to "spam email classification." Binary classifiers that classify emails into spam or ham were developed in the studies. The second highest number of articles is on the "multi-folder categorization of emails" (20 published articles), in which researchers developed a multiclass classifier that categorizes emails into various user-defined email directories. The third highest number of articles is related to "phishing email classification" (nine published articles), in which researchers developed binary classifiers that categorize emails into phishing or ham [15]. Spam filtering, however, does make things easier for the users. Thus, an email system that gives better outcomes consistently needs to be developed. This study develops a system that can be customized and tweaked for each user based on their individual preferences.

10.6 Gap Analysis

Spammers and bots constantly thrive when it comes to developing newer techniques and searching for loopholes in a spam classification system, and hence systems should be competent enough to handle newer techniques used by spammers and a variety of use cases. Many of the previously developed systems were combating only a few of these use cases. Despite the impressive accuracy obtained by such systems, they did not quite provide a comprehensive solution to the problem. A few systems first calculated a threshold value based on various keywords and parameters found in emails. Emails with a value more than the threshold value were classified as spam. However, the concept of the importance of an email is subjective.

One user might find emails with receipts crucial, whereas another user might consider it a waste of time. Hence, classification based on a specific threshold value would not work for all users at the same level. User context is an important factor that needs to be considered by the system. Some systems could achieve excellent accuracy in their outputs by implementing weighted or statistical probabilistic models. However, the system could not be deployed for the use of many users, thus limiting their scopes for research purposes only. Thus, these systems lacked real-world usage and vital feedback could not be obtained. Although research was carried out on trained email datasets, their application on continuous email management has not been active.

Third-party email clients have been popular with mobile users ever since email service providers have released application program interfaces for email management. Most users migrate to these email apps because the inbuilt email application by any provider has limited functionality. For example, Gmail included multi-account support in their default app a few years back, whereas third-party email clients already had included that feature long back. There are no email clients in the Android or iOS market that focus on email prioritization or spam classification. Most provide alternative features such as tracking opens, multi-account support, themes, and calendar integration but not ones that provide an email management experience at the initial level. This method aims to provide such an application to clear the email overload and have a smarter inbox experience with machine learning built into the app whose model can also be deployed as a middleware service.

10.7 Proposed System Architecture

Multiple machine learning models, supervised or unsupervised, can be trained and tested on the proposed model. Before committing an algorithm to the final model, it is necessary to compare the outcomes of various possible algorithms, such that the best one for a problem can be applied. The main goal here, very generally speaking, is to display important emails together and hide the unimportant ones. It is possible, but not advised, to manually block every sender since there's simply so many of them, and others that are sure to keep emailing any account. The model takes into consideration the following parameters: To field, From field, Message-ID, Cc/Bcc field, and so on in the email header. The email body with commonly used keywords and punctuations is also used for processing. Email data also requires preprocessing for it is preprocessed for machine learning. [16] Preprocessing is the primary step in data mining. In real-world cases, most of the data are not complete and contains incorrect values, missing values, and so on. The accuracy of classification or mining depends on the data being used. So, the first and foremost step to be performed before the mining task is to preprocess the data.

The model in Figure 10.1 is incorporated as a multistep process, which is carried out before the actual deployment of the model. The data required to train machine learning models are huge; as more and more data are fed into the models, the accuracy of the same keeps on increasing. This method proposes a workflow that helps in the aggregation of data used to train the machine learning model as well as to process the user's emails efficiently. Huge amounts of text data, such as email data, need a lot of data cleaning before they can be fed to the different models for testing. As there is no character limit for emails, the texts can be repetitive, garbled, and rife with punctuation and grammatical mistakes. This method proposes a way to clean and process the email data and make it ready for training and testing on different machine learning models to compare their results.

The proposed system architecture in Figure 10.1 focuses on retrieving emails through available third-party APIs and processes them appropriately for machine learning. Machine learning focuses on learning through previous inputs, and the data used for training and testing must follow a common format, without which the machine learning algorithms will fail

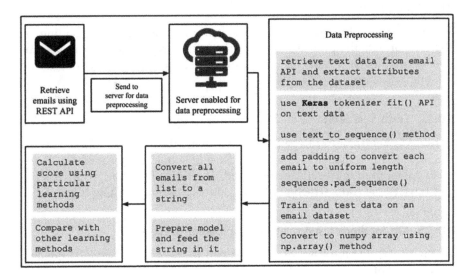

FIGURE 10.1
Proposed system architecture.

to give an ideal output despite being optimized thoroughly. The email data used for training and testing must have the same attributes such as timestamp, subject, title, body, whether the email is forwarded or not, and so on. This architecture focuses on collecting email data for training and testing in a common format such that the scores of various machine learning techniques used on a common dataset can produce accurate results through the various processing steps that generally a model of machine learning goes through. These steps are generally dependent on each other and aim to optimize the model.

10.7.1 Retrieving Emails

The first phase involves obtaining the data. The data for the model can be either preprocessed or raw data. Here, we consider the raw data. Generally, for machine learning models, the more the data, the better it is. The email dataset used for training must have the same attributes as the email that is retrieved using email APIs.

10.7.2 Sending Data for Preprocessing

Machine learning is a computationally heavy task, which is both time-consuming and inefficient on a personal computer; hence, the data is sent to a server for training, which has access to professional GPUs and other resources.

10.7.3 Preprocessing

Preprocessing helps make the data suitable for the model depending on the type of data the preprocessing works upon. Preprocessing is more time consuming for text data as the sentence semantics, punctuations, and annotations should be considered.

10.7.4 Train/Test Split

A Keras Tokenizer API can be used for verifying the train test split to varying degrees. The parameters for the train test split should be tweaked constantly so that the optimum results are obtained. The Keras Tokenizer API works on the TensorFlow code, which is an open source machine learning library.

10.7.5 API Methods Used for Processing Email Dataset

fit_on_texts(): fits the Keras Tokenizer API to the data to basically create a vocabulary of selected words.
text_to_sequence(): conserves the texts to a very broad and easy definition.
sequence.pad_sequence(): adds padding bits by default, which is done as pre-padding.
These methods help generalize the data as per the parameters of our algorithms.

10.7.6 Creation of Model and Training

Here, the preprocessed data is ready to be fed into the model and train it. It is very important that the training and testing data are partitioned well so that the data testing set is never visible to the model. This enables a higher score without any underlying bias from the algorithm.

10.7.7 Verification

After training a particular model, it should be verified on the testing data, following which its score should be compared with the results of other machine learning algorithms and the parameters should be adjusted accordingly for the most optimum score. The testing and training data should be adjusted for more favorable results.

10.8 Retrieval of Email Data for Spam Classification

Free email service providers such as Gmail, Yahoo! Mail, Hotmail, and Outlook are used by millions of people worldwide. With Gmail alone having 1 billion active users and Hotmail having 400 million users, these providers contribute a huge chunk to email management. With the services of these email providers used for enterprise solutions, these companies have released their own email APIs to integrate into devices for email management. The APIs allow the authenticated user to read and view emails, along with modifying the labels and accessing these emails from another email client. For accounts with custom domains apart from the services provided by Microsoft Outlook, the JavaMail API by Oracle provides a platform for third-party email management. This API can be called from any Android or iOS device, and the programmer can implement any custom functionality in the email client without compromising on the security of the emails. The JavaMail API provides a protocol- and platform-independent framework to send and receive emails. The different protocols supported and used in JavaMail API are SMTP, POP, and IMAP.

JavaMail API can send or receive emails regardless of the implementation of a single specific protocol because of the various protocols being implemented presently. Along with this, the API can fetch emails from all the email service providers be it Google, Yahoo, Hotmail, and so on. This ensures versatility to the approach for the retrieval of emails. The JavaMail API uses service provider interfaces (SPIs), which provide intermediary services to java application with which it can deal with different protocols [18].

10.9 Results and Discussion

The dataset used for testing different machine learning algorithms is the Ling Spam dataset. For classifiers that process text data, a dataset focusing on linguistics is an ideal one. The number of ham emails (legitimate emails) should also be greater than that of spam emails so that the training data can not only provide good accuracy for incoming spam emails but also avoid false positives. The chosen dataset, that is, the Ling Spam dataset, comprises 2412 ham emails obtained from referring the digests, and 481 of the emails are classified as spam, which was retrieved from an author of the entire email corpus. The dataset is situated in folders with each email in an individual text file, with the subject and the body of the email specified under individual labels. After preprocessing the Ling Spam dataset using count vectorization in python, the features were extracted from the emails and fit transform was used to map the data on a word-to-integer basis. The labels were assigned to the emails based on the parameters and features, which was further used for training. The trained data was tested upon along with the count vectorization object made from stop words and maximum features in the email dataset. A comparative study of the algorithms that were implemented is discussed next. The number in each individual column represents the accuracy or the score. In machine learning, score represents how an algorithm performs on various parameters. Scores may refer to a quantification of the performance on various metrics. The choice of metrics is highly correlated to the parameters that are to be optimized for a given problem.

The algorithms compared in Figure 10.2, used to analyze the performance against a common email dataset, are supervised machine learning algorithms. These algorithms were initially trained using python libraries like numpy, scikit, and TensorFlow. A common dataset was used so the comparative performance could be assessed correctly. After obtaining the results, factors such as future scalability, overall performance, number of features, and labeling were considered to choose an optimum algorithm for email spam classification. The main factor to take into consideration was the scalability of an algorithm as some algorithms provide excellent accuracy but show poor performance for a large dataset or when many features are considered. Although there is no ideal or perfect algorithm for email spam classification, these important factors should be considered while developing a third-party email management application for deployment.

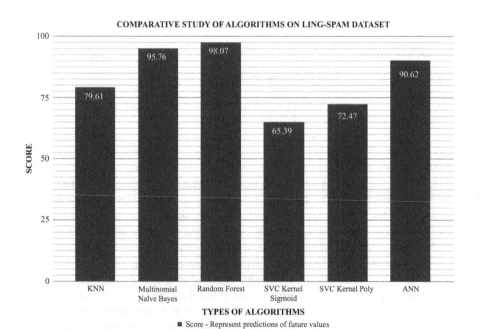

FIGURE 10.2
Comparative analysis of machine learning algorithms.

10.9.1 k-Nearest Neighbor

The k-nearest neighbors (KNN) algorithm is one of the simplest to implement in the machine learning field and can be used for classification and regression. However, currently, it is predominantly used for classification [19] [20]. The distance between the nodes can be calculated using Euclidean, Manhattan, or minkowski distance depending on the dimensionality. All the three distance formulas are valid for continuous variables; however, for categorical variables, hamming distance must be used. The optimal value can be inferred only after multiple plotting and considering the error rate for each value.

As all the training examples are stored in memory, this technique is also referred to as a memory-based classifier. Another problem of the presented algorithm is that there seems to be no parameter that can be tuned to reduce the number of false positives. This problem however can be easily solved by changing the classification rule to the following l/k-rule: if l or more messages among the KNN of x are spam, classify x as spam; otherwise, classify it as legitimate mail [21]–[35].

10.9.2 Decision Tree

The decision tree is a supervised machine learning algorithm. As the name suggests, decision tree is a tree-like structure having nodes and branches. Each branch has a test condition that is tested on the immediate node. At every state, the purest form of data needs to be determined, that is, the data that gives us the maximum information for which many implementations can be used, such as "Gini Index," "Chi-Square," and so on. Although this algorithm can be used for both classification and continuous outputs, currently it is mainly used for classification.

10.9.3 Multinomial Naive Bayes (NB)

NB is a machine learning algorithm that works using the conditional probability distribution model. The multinomial NB is a type of NB, and it states whether a sample has multinomial distribution. Multinomial NB works well when used with data whose counts can be easily obtained such as in texts.

10.9.4 Random Forest

Random forest is a supervised machine learning classification algorithm. The difference between random forest algorithm and the decision tree algorithm is that in random forest, the processes of finding the root node and splitting the feature nodes will run randomly. The tree in the random forest is randomly created according to the features compared to the decision tree model, where the programmer assigns the features. [36] The random forest has an advantage in scaling as it avoids the overfitting problem, which occurs commonly in machine learning where the training phase learns the noise in the test dataset such that it affects the performance negatively. Random forest is highly useful for feature classification and extraction.

10.9.5 SVC Kernel

The support vector method plots the data available in an n-dimensional space, where n is the number of features available. The support vector then classifies the two classes with respect to a hyperplane. This is implemented using the sci-kit learn library in python. The main advantage of this method is that it performs well in high dimensional data, but training is expensive as it does not process data noise effectively.

10.9.6 Artificial Neural Networks

The working of the artificial neural network is inspired by the working of neurons in a human brain. The basic structure of the ANN tries to mimic the workings of the human brain, like the thought process that occurs when neurons are fired in the brain. The artificial neural network consists mainly of three layers—an input layer, a hidden layer, and an output layer. In the case of email spam classification, the input consists of nodes, with each feature of the email body as a node such as Cc, Bcc, the frequency of emails sent, email body text, and so on. The hidden layers have a transient form, which emulates a probabilistic behavior. Each hidden layer tries to predict the activation rate for the next hidden layer while finally entering the output state. This is a model that performs better as more data is made available for training.

The complete overview and observations of the algorithms used for testing on the Ling Spam dataset are summarized in Table 10.1.

10.10 Serving Machine Learning Models

Once the machine learning model is properly trained, the next step is exporting the model to the necessary platform for implementation purposes. This task is called serving. Figure 10.3 shows a workflow by which the algorithm can be easily deployed on other platforms necessary to apply the output provided by the model for real-world applications. Deploying the machine learning models in production is a crucial part of a model and is incomplete without a proper deployment in production.

Machine learning models were traditionally deployed on servers as they consisted of GPUs capable of processing huge amounts of data for a

TABLE 10.1

ALGORITHM	TYPE	STRUCTURE	ADVANTAGE	DISADVANTAGE
k-Nearest Neighbor	Supervised	Distance plotting	Higher accuracy for classification	Results dependent on value of k
Decision Tree	Supervised/ unsupervised	Tree structure	Requires less data cleaning	Overfitting of data occurs, leading to poor accuracy
Multinomial Naive Bayes	Supervised	Conditional probability	Provides an accurate classification	Does not work well on large datasets
Random Forest	Supervised	Tree structure	Achieves higher scores on small datasets	Works poorly on high-dimensional data
Support Vector Classification Kernel	Supervised	Decision boundary	Memory efficient, works in high-dimensional spaces	Training time is higher
Artificial Neural Networks	Supervised	Nonlinear functional approximation	Scales excellently, number of nodes can be changed	Needs a lot of training data for higher score

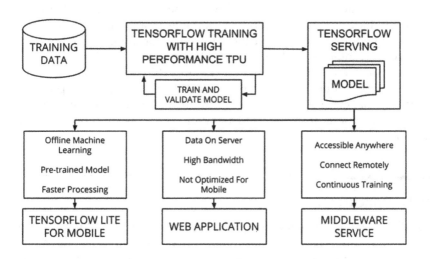

FIGURE 10.3
An example of serving machine learning models using TensorFlow.

specific purpose or as a middleware service deployed as a SaaS. Serving models on servers were cumbersome and expensive and their capabilities were limited to enterprises. Mobile users did not have access to easy machine learning capabilities as a mobile device has limited processing power, bandwidth, battery, and storage. With the rise of mobile users, researchers started looking for a mobile-first machine learning solution. In late 2017, machine learning models could be served on devices that use low computing power, can work offline without the data embedded, and can be small (up to 2MB) without the extreme power requirements of a GPU. As Figure 10.3 specifies, TensorFlow Lite, which is an open source solution for machine learning, uses many techniques for achieving low latency such as optimizing the kernels for mobile apps, pre-fused activations, and quantized kernels that allow smaller and faster (fixed-point math) models. [37] It specifically provides and interfaces for hardware acceleration on hardware devices such as Android, iOS, and Raspberry Pi. Machine learning for mobile devices primarily uses a file structure called Flat Buffers, which represents hierarchical data while supporting multiple and complex data structures such as scalars, arrays, extensible array, balanced tree, heap, and so on with a complete backward compatibility in a way that it can still be computed directly, without the data being parsed or unpacked. This helps machine learning operations work at an optimum level in a low-powered device without consuming extreme resources.

Figure 10.3 shows the trained TensorFlow model, which is frozen by converting the dynamic variables into static and embedded into an iOS or Linux device, which is called Java API. Java API is a wrapper around the C++ API, which invokes the device's kernel using a set of kernels that performs hardware acceleration. A portable mobile-first solution can be implemented for serving machine learning for offline interaction, low latency, and interest in stronger user data privacy paradigms where user data do not need to leave the mobile device. This mobile solution, coupled with serving models from servers with high computing needs, provides an open and flexible machine learning production and deployment environment to promote research, collaboration, and distribution of machine learning.

10.11 Conclusion

As discussed above, although email spam might look like a harmless distraction, it is actually a dangerous facade. Most spam emails might be benign, but even the 1% that have been specially created with malicious intents could prove severe. Whaling, spear phishing, website forgery, and clonephishing can be included in this category. Classification of emails is thus important for the security of not just the user, but also the email service provider, as spam emails still consist of a large chunk of emails sent daily, which add up to a large amount of server storage space, resulting in limited storage and bandwidth for the administrator. The algorithms mentioned above provide us with different results depending on the dataset used, the size of the dataset, and the unique advantages and disadvantages of each. In any problem, it is important to compare the workings of every proposed solution. The biggest reason for such a comparison is that a user can use any of the above-mentioned algorithms, from artificial neural networks to random forests, completely based on their requirements at the given time. These comparisons show certain biases that might have been missed when using a single approach, and they provide developers with a certain flexibility to use any of them at their own will.

The easy availability of machine learning libraries has allowed people from all over the world to implement these algorithms in various ways. This has led to a huge increase in popularity for not just python as a programming language, but aspects of machine learning and algorithms in general. TensorFlow already has hundreds of thousands of users because it is easily accessible and easy to learn. One potential application using the above-mentioned machine learning algorithms is email classification. Despite having classified emails as spam and ham, many users still face the problem of email overload because of multiple accounts, frequency of emails, and even job designation. One further scope that can be extended from spam classification is prioritizing the ham emails between the tags of "Priority" and "Other." It is possible that important emails such as meeting reminders, deadlines, delivery status, and so on might get overlooked as the user's inbox is constantly swamped with emails.

The future scope of this spam classification involves prioritization of the already-classified ham emails under parameters such as domain, frequency of emails previously sent by that user, Cc/Bcc, and specific keywords in the body of the email, which would act as inputs to a neural network model, thereby classifying each message as either high priority (if it requires an urgent reply) or normal to low priority. This would provide a better overall email management experience to the user. The model currently implements a neural network (whichever is to be used) with a predefined and labeled dataset. When it is implemented on a user's inbox, it classifies each incoming email as spam or non-spam automatically and accordingly displays it only in the required section or folder. Not only are users shielded from unnecessary emails, saving precious time, they are also protected from potentially harmful ones, and thus the benefits are twofold. As attacks and threats by emails become more versatile in their methods, this elementary measure gives us a good starting point to provide both a secure and productive environment to the user.

References

[1] Ramzan, Zulfikar (2010). "Phishing attacks and countermeasures". In Stamp, Mark & Stavroulakis, Peter eds. Handbook of Information and Communication Security. Springer. ISBN 9783642041174

[2] Samuel, Arthur (1959). "Some Studies in Machine Learning Using the Game of Checkers". IBM Journal of Research and Development.

[3] Top 15 Python Libraries for Data Science in 2017 [https://medium.com/activewizards-machine-learning-company/top-15-python-libraries-for-data-science-in-in-2017-ab61b4f9b4a7]

[4] Kuldeep Yadav, Ponnurangam Kumaraguru, Atul Goyal, Ashish Gupta and Vinayak Naik. SMS Assassin: Crowdsourcing Driven Mobile-based System for SMS Spam Filtering Copyright 2011 ACM. 978-1-4503-0649-2 $10.00

[5] Lee Sproull and Sara Kiesler. (1986, Nov). Reducing social context cues: Electronic mail in organizational communication. Management Science (32, 11)

[6] Mackay, W. Diversity in the use of electronic mail: A preliminary inquiry. ACM Transactions on Office Information Systems 6, 4 (1988), 380–397

[7] Sahami, M., Dumais, S., Heckerman, D., & Horvitz, E. (1998, July). A Bayesian approach to filtering junk e-mail. In Learning for Text Categorization: Papers from the 1998 workshop (Vol. 62, pp. 98-105)

[8] Zhang, L., Zhu, J., & Yao, T. (2004). An evaluation of statistical spam filtering techniques. ACM Transactions on Asian Language Information Processing (TALIP), 3(4),243-269.

[9] Boykin, P. O., & Roychowdhury, V. P. (2005). Leveraging social networks to fight spam. Computer, 38(4),61-68

[10] Wittel, G. L., & Wu, S. F. (2004, July). On Attacking Statistical Spam Filters. In CEAS.

[11] Tretyakov, K. (2004, May). Machine learning techniques in spam filtering. In Data Mining Problem-oriented Seminar, MTAT (Vol. 3, No. 177, pp. 60-79)

[12] Alan Gray Mads Haahr. Personalised, Collaborative Spam Filtering. Enterprise Ireland under grant no. CFTD/03/219

[13] Goodman, J., Cormack, G. V., & Heckerman, D. (2007). Spam and the ongoing battle for the inbox. Communications of the ACM, 50(2),24–33

[14] Satyajeet Raje, Sanket Tulangekar, Rajshekhar Waghe, Rohit Pathak, Parikshit Mahalle. Extraction of Key Phrases from Document using Statistical and Linguistic analysis. Proceedings of 2009 4th International Conference on Computer Science & Education. 978-1-4244-3521-0/09/$25.00 ©2009 IEEE

[15] Ghulam Mujtaba, Liyana Shuib, Ram Gopal Raj, Nahdia Majeed, Mohammed Ali Al-Garadi. Email Classification Research Trends: Review and Open Issues. 2169-3536 (c) 2016 IEEE

[16] Anju Radhakrishnan et al. Email Classification Using Machine Learning Algorithms. International Journal of Engineering and Technology (IJET) Vol 9 No 2 Apr-May 2017. DOI: 10.21817/ijet/2017/v9i1/170902310.

[17] Jim Keogh. J2EE: The Complete Reference, 2002

[18] Khorsi. "An overview of content-based spam filtering techniques", Informatica, 2007

[19] W.A. Awad, S.M. ELseuofi. Machine Learning Methods For Spam E-Mail Classification. International Journal of Computer Science & Information Technology (IJCSIT), Vol 3, No 1, Feb 2011

[20] How Random Forest Algorithm Works in Machine Learning https://medium.com/@Synced/how-random-forest-algorithm-works-in-machine-learning-3c0fe15b6674

[21] Wang, Cunlei, Zairan Li, Nilanjan Dey, Amira Ashour, Simon Fong, R. Simon Sherratt, Lijun Wu, and Fuqian Shi. "Histogram of oriented gradient based

plantar pressure image feature extraction and classification employing fuzzy support vector machine". Journal of Medical Imaging and Health Informatics (2017).

[22] Zemmal, Nawel, Nabiha Azizi, Nilanjan Dey, and Mokhtar Sellami. "Adaptive semi supervised support vector machine semi supervised learning with features cooperation for breast cancer classification". Journal of Medical Imaging and Health Informatics 6, no. 1 (2016): 53-62.

[23] Wang, Yu, Fuqian Shi, Luying Cao, Nilanjan Dey, Qun Wu, Amira S. Ashour, Simon Sherratt, V. Rajinikanth, and Lijun Wu. "Morphological segmentation analysis and texture-based support vector machines classification on mice liver fibrosis microscopic images". Current Bioinformatics (2018).

[24] Ahmed, Sk Saddam, Nilanjan Dey, Amira S. Ashour, Dimitra Sifaki-Pistolla, Dana Bălas-Timar, Valentina E. Balas, and João Manuel RS Tavares. "Effect of fuzzy partitioning in Crohn's disease classification: a neuro-fuzzy-based approach". Medical & biological engineering & computing 55, no. 1 (2017): 101-115.

[25] Cheriguene, Soraya, Nabiha Azizi, Nilanjan Dey, Amira S. Ashour, Corina A. Mnerie, Teodora Olariu, and Fuqian Shi. "Classifier Ensemble Selection Based on mRMR Algorithm and Diversity Measures: An Application of Medical Data Classification". In International Workshop Soft Computing Applications, pp. 375-384. Springer, Cham, 2016.

[26] Zemmal, Nawel, Nabiha Azizi, Nilanjan Dey, and Mokhtar Sellami. "Adaptative S3VM semi supervised learning with features cooperation for breast cancer classification". Journal of Medical Imaging and Health Informatics 6, no. 4 (2016): 957-967.

[27] Dey, Nilanjan, Amira S. Ashour, Sayan Chakraborty, Sourav Samanta, Dimitra Sifaki-Pistolla, Ahmed S. Ashour, Dac-Nhuong Le, and Gia Nhu Nguyen. "Healthy and unhealthy rat hippocampus cells classification: a neural based automated system for Alzheimer disease classification". Journal of Advanced Microscopy Research 11, no. 1 (2016): 1-10.

[28] Kamal, Md Sarwar, Mohammad Ibrahim Khan, Kaushik Dev, Linkon Chowdhury, and Nilanjan Dey. "An Optimized Graph-Based Metagenomic Gene Classification Approach: Metagenomic Gene Analysis". In Classification and Clustering in Biomedical Signal Processing, pp.290-314. IGI Global, 2016.

[29] Dey, Nilanjan, ed. Classification and clustering in biomedical signal processing. IGI Global, 2016.

[30] Bhattacherjee, Aindrila, Sourav Roy, Sneha Paul, Payel Roy, Noreen Kausar, and Nilanjan Dey. "Classification approach for breast cancer detection using back propagation neural network: a study". Biomedical image analysis and mining techniques for improved health outcomes (2015): 210

[31] Maji, Prasenjit, Souvik Chatterjee, Sayan Chakraborty, Noreen Kausar, Sourav Samanta, and Nilanjan Dey. "Effect of Euler number as a feature in gender recognition system from offline handwritten signature using neural networks". In Computing for Sustainable Global Development (INDIACom), 2015 2nd International Conference on, pp. 1869-1873. IEEE, 2015.

[32] Kotyk, Taras, Amira S. Ashour, Sayan Chakraborty, Nilanjan Dey, and Valentina E. Balas. "Apoptosis analysis in classification paradigm: a neural network based approach". In Healthy World Conference, pp. 17-22. 2015.

[33] Samanta, Sourav, Sk Saddam Ahmed, Mohammed Abdel-Megeed M. Salem, Siddhartha Sankar Nath, Nilanjan Dey, and Sheli Sinha Chowdhury. "Haralick features based automated glaucoma classification using back propagation neural network". In Proceedings of the 3rd International Conference on Frontiers of Intelligent Computing: Theory and Applications (FICTA) 2014, pp. 351-358. Springer, Cham, 2015.

[34] Chatterjee, Sankhadeep, Sarbartha Sarkar, Sirshendu Hore, Nilanjan Dey, Amira S. Ashour, Fuqian Shi, and Dac-Nhuong Le. "Structural failure classification for reinforced concrete buildings using trained neural network based multi-objective genetic algorithm". Structural Engineering and Mechanics 63, no. 4 (2017): 429-438.

[35] Chatterjee, Sankhadeep, Sarbartha Sarkar, Sirshendu Hore, Nilanjan Dey, Amira S. Ashour, and Valentina E. Balas. "Particle swarm optimization trained neural network for structural failure prediction of multistoried RC buildings". Neural Computing and Applications 28, no. 8 (2017): 2005-2016

[36] How to deploy Machine Learning models with TensorFlow https://towards datascience.com/how-to-deploy-machine-learning-models-with-tensorflow-part-1-make-your-model-ready-for-serving-776a14ec3198

[37] TensorFlow: Large-Scale Machine Learning on Heterogeneous Distributed Systems (Preliminary White Paper, November 9, 2015)

[38] Yoo, S. (2010). Machine learning methods for personalized email prioritization. PhD. Carnegie Mellon University.

[39] Alurkar, A., Ranade, S., Joshi, S., Ranade, S., Sonewar, P., Mahalle, P. and Deshpande, A. (2017). A proposed data science approach for email spam classification using machine learning techniques. In: 2017 Internet of Things Business Models, Users, and Networks. [online] Copenhagen: IEEE. Electronic ISBN: 978-1-5386-3197-3, IEEE Catalog number: CFP17M58-ART Available at: http://ieeexplore.ieee.org/document/8260935/ [Accessed 18 Jan. 2018]

11

Malware Prevention and Detection System for Smartphone

A Machine Learning Approach

Sachin M. Kolekar
Department Computer Engineering, Zeal College of Engineering and Research, Pune, Maharashtra, India

CONTENTS

11.1 Introduction

Number of Smartphone users has increased tremendously in last decade, and it continues to increase every day. This in turn attracts malware developers to target smartphones and perform their malicious activities. Nearly all of these Intrusion Detection Systems are demeanor-predicated, for example, they don't plan on a record of maleficent code design, as in the case of signature-established Intrusion Detection System. In this chapter, we report machine learning–predicated malware diagnosis system for android smartphone users that uses battery life monitoring, and malware uncovering techniques. This system exploits machine learning techniques.

An Android in service system is known to be the most admired and widely used operating system. According to the Gartner story, Android has subjected in use system market by capturing 81.7% of total market share by the end of 2016. Android is the most dominant and widely used OS in the market. With the speedily mounting status of the android OS, the expansion of malicious application conjointly exaggerated. The android platform offers subtle functionalities at terribly low price and has become the foremost well-liked OS for hand-held devices. In official android market there are many applications that are being downloaded by the users in an exceedingly sizable amount on a daily basis. As android permits downloading and putting in an application from third-party marketplaces; malware developers take advantage of this by repackaging and uploading a notable application from Google Play Store. Malware includes PC viruses, Trojan horses, ad-ware, backdoor, spy-ware, and alternative malicious programs, which are designed to harm the OS and steal user data. To stop such malware, it is necessary to possess a correct and deep understanding of them along with security measures to guard user information that might be taken consequently.

11.2 Related Works

We have referred to the Crow droid that mimics Trojan-like malware, which upon victimization of an android operating system of good phones, analyzes the range of clocks of every system call that has been issued by an application throughout the implementation. The actual claim differs from its trojanized version since its problems vary over a distinct range of system calls. The Crow droid assists in building a vector of m features. Another IDS that depends on machine learning expertise is Andromaly [3], which monitors the smartphone and user behaviors based on a number of parameters, spanning from sensor activities to CPU utilization. Eighty-eight options are acclimated to explain these demeanors. The authors claim to find abnormalities equivalent to a Multi-Level Anomaly Detector [5] using thirteen features to find android malware for each kernel level and utilize level. MADAM has been tested on authentic malware beginning with the untamed ecumenical-observe approach which is able to find malware contained in undisclosed state. It monitors smart phones to extract options that may be utilized by machine learning techniques and uses Kirin security to find the anomalies. The framework consists of an observation client, a Remote Anomaly Detection System

(RADS) and a conjure element [7]. RADS is an internet-based service that receives data from the observation of a client and the observed features are then maintained in a database and used to implement machine learning techniques and Kirin security. In [9], the authors propose a demeanor-predicated malware discovering system (pBMDS) that maps user's inputs with system calls to notice irregular activities cognate to SMS/MMS. It suggests Kirin security accommodation for android that performs lightweight certification of applications to weaken malware at the time of installation. Kirin certification uses security rules and matches the undesirable properties in security design bundled with the applications.

11.3 Proposed Work

For every Android claim, we retrieve several selected skin tones from the equivalent application package (APK) file [2]. In addition, permission from the packages are extracted from Android manifest.xml file. The standards of selected skin tone are stored as a binary number, which is pictured as a chain of comma not speaking values, where every sequence includes the name of an attribute improvement, the information variety of the feature, and information of the feature. Sanctions on paper might avail users or reviewers establish malware, if the sanction requirements of baneful applications vary from permission requests of mundane, non-malevolent applications. Our goal is to see whether permissions are a subsidiary implementation for identifying malware in practice. This section evaluates whether the Android malware in our data set exhibits anomalous patterns of sanction requests. Could a simple classifier utilize the sanction needs of malware to identify maleficent from non-maleficent claim? We opted to study Android malware because Android has the most widespread sanction system and therefore represents the best display place for sanction-predicated classifier action. To perform the study, we collected the sanction requests made by each piece of Android malware. This data was not always accessible from anti-virus researchers, but we were able to identify the sanction requisites for 11 of the 18 Android malware applications[8]. The sample skin tones are explained in the next list.

1. **android.permission.INTERNET**: This permission is required for the application that needs to connect to the internet. When user gives permission to the Internet, the application can fetch or upload data to its server. Without this permission, applications cannot access data because some functionalities of the application may not work properly[2].

2. **android.permission.CHANGE_CONFIGURATION**: This feature allows users to change the configuration of the application. Configuration

of the application may change by its state. This application allows changing the current configuration of an application.

3. **android.permission.WRITE_SMS**: With the help of this, a go-ahead claim can access the message handler of the devices, which will be used to write a message [2].

4. **android.permission.SEND_SMS**: This permission allows to send message from client side to server side.

5. **android.permission.CALL_PHONE**: This permission allows application to initiate phone calls without going through the dialer, as the applications may use their own dialer[2].

11.4 Implementation Details

In this chapter we implemented six different parameters for malware on android platform. These malwares can steal the personal information of the users such as serial numbers of the software. Different malwares like ransom ware can bypass the device security to lock down the whole system. As we are using six different parameters, various malware profiles of the same family can be engendered. The flow shown in Figure 11.1 describes how a malware is detected. Moreover, the parameters themselves can be randomized rather than being a fine-tuned value. Using random six parameters, intriguing malware profiles can be produced for further analysis.

There may be a hidden call recorder in many malicious applications, and whenever a user is on a phone call this malicious application will eavesdrop the whole conversation of the user. These recorded files are then stored in a local database on the phone and whenever the user connects to the internet this application can send all the recorded files on their server with the help of internet permission. This malware may include the following parameters:

1. ❼**MAX_DURATION**—This will record the duration of recorded calls.

2. ❼**MAX _FILESIZE**—Actual volume of the recorded phone calls.

3. ❼**NUM_SKIPPED_CALLS**—identifies that each of the (n+1)phone calls are completely recorded, where n represents the value of NUM_SKIPPED_CALLS.

4. ❼**INTERVAL_RECORD**— This includes the total recorded duration of a phone call.

5. ❼**SHOULD_UPLOAD**—All recorded contents can be uploaded on server side.

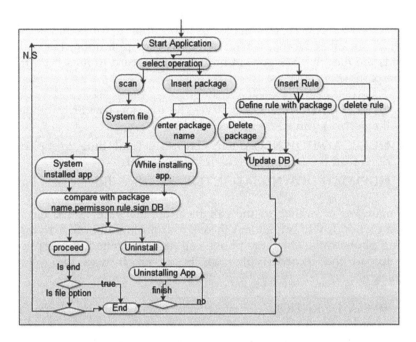

FIGURE 11.1
Malware tracking and detection system flowchart.

6. ❼**DELETE_LOCAL**—Unnecessary files which are in local database will be removed.

In Denial of service malware attack, the network is flooded by the unnecessary services and blocks other legitimate services. This malware uses the threads to flood the whole system and, as these threads have been used by the malicious application, whenever the user tries to place a call, the phone hangs. When the number of gearspawn is above two hundred, the phone hangs and goes into rebooting mode. Thus, this type of malware paralyzes the device by loading the central processing unit beyond its limit. It uses certain parameters such as:

1. ❼**MAX_THREADS**—It shows the maximum number of threads spawn by a malware.
2. ❼**NUM_MULTIPLICATIONS**—maximum number of multiplications performed by each and every thread.
3. ❼**INTERVAL_RESTART**—duration after which all the spawned threads are destroy before new ones are spawned.
4. ❼**INTERVAL_SLEEP**—A specific length of time for which the malware is in sleeping mode before new gears are being spawn.

This type of malware is usually designed to steal the personal information of the user. This type of malicious application uploads all the data to its server. In other words, when user starts the application, the malware uploads the files that are present in the memory card to its server. It uses certain parameters such as:

1. ❼UPLOAD/DOWNLOAD_BW—The total bandwidth provided by the network can be monitored.
2. ❼UPLOAD/DOWNLOAD_INTERVAL—Total time to upload or download the local files to or from the server.
3. ❼UPLOAD/DOWNLOAD INTER_LIM—Specifies the data limit.

This malware is similar to the call recorder malware as we have discussed earlier. It will implement the eavesdropping on both arriving and retiring phone calls. Once the phone call has terminated, the application will upload the recorded files on its server. It uses the following parameters.

1. ❼MAX_DURATION—Greatest time extent that the phone call is recorded mode.
2. ❼MAX_FILE_SIZE—Greatest file size of the recorded call.
3. ❼INTERVAL_RECORD—Length of each recording mode through a phone call
4. ❼INTERVAL_SLEEP—Duration for which the malware goes to sleep mode.

Spy camera- is a type of the malware in which malicious application gets the permission of camera from the user. This malware can access the camera whenever application wants without notifying the user. It can take the pictures as well as videos of the user and can upload it on their server. It uses the following parameters:

1. ❼SNAP_INTERVAL—A snap is taken from the mobile devices' camera; all data or snap shots are stored locally on the phone.
2. ❼PIC_DSAMPLE_RATIO—Specifies the Down Sample ratio.
3. ❼PIC_COMP_QUALITY—Specifies the amount for reducing the raw image saving the image.
4. ❼SHOULD_UPLOAD—Uploading the complete picture to a server.
5. ❼DELETE_LOCAL—Deleting all local copies of the pictures taken.

Spy recorder is a type of malware which uses the phone microphone remotely. A malicious application gets the microphone's permission from the user to turn on or off the microphone without user's consent. Recorded calls can be uploaded to the server:

1. ❼MAX_DURATION—Greatest instance for which the talk is recorded.
2. ❼MAX_FILE_SIZE—Greatest size or space of the recorded file.
3. ❼INTERVAL_RECORD—Length of every recording during a phone call
4. ❼INTERVAL_SLEEP—Duration for which the malware is in sleep mode.
5. ❼SHOULD_UPLOAD—Uploading recorded talk to a server.
6. ❼DELETE_LOCAL—Deleting or destroying all the local copies of the recorded conversation.

Up to this point, we have discussed six different parameters used by the malicious application. Mostly a malicious application uses these parameters to achieve its malicious intent. Our system can easily detect the malicious activity with the help of these parameters. We can also add some data mining approaches to find malwares such as footprint matching with the help of behavior based applications.

11.5 Steps of Smartphone Security

As we have seen in the Introduction that the number of users of smartphones has increased in recent times, threats to the software have also increased along with it. So, it is very necessary to secure our devices properly. Users can secure their devices with the help of the following techniques [4]:

1. **Set PINs and password**: A malware can easily access the applications, which are installed in our system without notifying the user. So, setting the pin and password for applications allows user to safely authenticate the system. Devices should be secured with password.
2. **Do not modify your Smartphone security settings**: As android allows rooting privileges, which allows modifying a system, users should not perform the rooting or jailbreaking or the devices as it leads to reduced security of devices.
3. **Backup and secure your data**: Many malwares can delete the data, so it is always suggested to keep a backup of the data. If by any means the data is lost, we can restore the data anytime. Data can be stored on cloud, which can be fetched anytime.

4. **Only install apps from trusted sources**: There are many application stores available other than official stores. For example, on android there are many Chinese app stores that are flooded with malicious application. Therefore, user should download the application only from trusted sources such as Play Store for android and App Store for iOS.

5. **Set up security apps that enable isolated location and wiping**: It is advisable that there should be anti-viruses or antimalware to secure our devices. Installing such application increases the security of devices.

6. **Accept updates and patches to your Smartphone software**: Smartphone companies are also working to reduce the malwares. Therefore, user should keep their devices up to date. Whenever new patches are launched, the user should download it immediately.

7. **Be smart on open Wi-Fi networks**: Open wifi network can monitor the user's activity via smartphones which are connected to the wifi. Therefore, user should not connect to any open wifi to access internet. Wifi admin may monitor user authentication's process, so one should be always smart while accessing the public wifi.

8. **Wipe down data on your mature phone before you donate, resell, or recycle it**: When user is changing a device or selling it to other users, user should wipe out the original data before selling it.

9. **Report a stolen Smartphone**: When a device is lost, the user should always report it to the: Federal Communications Commission and also ask the network provider to shut down the sim card. This will notify all the chief wireless accommodation providers that the phone has been lost or stolen and will sanction for remote "bricking" of the phone so that it cannot be activated on any wireless network without your sanction at the time of installation of applications for malware detection.

11.6 Relevant Mathematical Model

11.6.1 Set Theory Analysis

a. Let 'S' be the | Malware deterrence and uncovering system using Smartphone as the final set

S = {C, P, M}

b. Identify the inputs as C, P and M

S = {C, P, M}
C = {C1, C2, C3, ...,Cn | 'C' gives the .apk file}

TABLE 11.1

Mathematical Model

Sr No.	Mathematical Model	Description	Observation
1	C = {C1, C2, C3, ..., Cn}	C gives .apk file	As the size of current inputs are variable we used SQlite DB
2	P = {P1, P2, P3, ..., Pn}	P gives installed files	As the size of previous inputs are variable we used SQlite DB
3	M = {M1, M2, ..., Mn}	M gives permission set	As the alert message is used ADT
4	O={O1, O2, O3,...,On}	'O' gives malware detection	

P= {P1, P2, P3, ...,Pn | 'P' gives installed file}
M= {M1, M2, M3 | 'M' gives permission set}

  c. Identify the outputs as O

S = {C, P, M...
O = {O1, O2, O3, On | 'O' is the malware detection}

  d. Identify the functions as 'F'

F = {scanning byPackage (), Scanning byPermission (), scanning bySignature ()}
F1 =Scanningbypackage(C)=O'::take .apk file as input.
F2 = Scanning ByPermission(P)=O':: take .apk file as input.
F3 = Scanning BySignature(C)=O':: take .apk file as inpit.

Data Structures Used
For 'C': XML, Java
For 'P': XML, Java
For 'O':XML, Java

11.7 Apparent Study for Malware Uncovering in Android (MAMA)

Mobile phones have had a great success in the market as they offer maximum services equivalent to that of a laptop or desktop computer. Besides, the number of applications available for Android mobile phones has also increased. Google provides programmers the chance to upload and vend applications in the Android market, but malware writers upload their maleficent code there. In light of this setting, we present here Analysis for Malware Detection in Android (MAMA), an incipient method that extracts several skin tones from the Android applications to construct machine learning classifiers and detect malware.

11.7.1 Materials and Methods

This section describes variants of the method for the detection of Android malware applications and the dataset utilized for the validation process.

11.7.2 Dataset Description

This section describes how the dataset has been compiled along with the features to be noted in compiling the dataset.

It must be sundry: It should show a variety in the applications that exists in the Android market.

It must be relative to the number of samples that previously exist of every category of application. To this end, two special datasets were created. First dataset is self-possessed of malicious software, while the second one is formed by benign applications.

11.7.3 Feature Engineering

In this section, a diverse type of feature sets has been observed, which has been utilized in order to detect Android malware. These features have been collected from Android Manifest.xml file, which is available within each and every Android application. The malware applications present a more rigorous utilization of the sanctions cognate with the transferring and reception of textual messages.

These attribute sets are:

(i) the authority needed for the claim, under the uses-permission tag and

(ii) the attribute under the uses-features group in the Android Manifest File. In order to acquire these features, we first extracted the sanctions utilized by each of the applications. To this extent, we employed the AAPI (Android Asset Packaging Implement), which exists within the set of implements provided by the Android SDK.

These features were culled for two main motives:

- The process of accumulating them has a low computing overhead.
- The different deportments that may be present within the application are manifested by them. In order to compare the feature sets, we have evaluated both feature sets discretely and, then, cumulated in a unique vector. We can withal realize that the differences between the number of maleficent applications. This type of application requires the INSTALL_PACKAGES. This sanction, cumulated with the Internet connection of the contrivance, may sanction the assailant to install any type of application. Other kind of sanctions, such as the ones

designated to accumulate users' data, have a high pertinence in the maleficent applications.

11.7.4 Authorizations of the Manifest File

In this section, we have performed section analysis, as it is requested by the applications in the dataset in order to quantify their pertinence in the malware detection process.

```
<uses-permission android:name="string"/>
```

Accordingly, there are numerous strings that are utilized for declaring the sanction utilization of the diverse Android applications, such as "android. permission.CAMERA" or "android.permission.SEND_SMS". Also, analysis is done for the number of sanctions and their frequency in a precedent step for kenning their administration within our dataset.

11.7.5 Other Aspects of the Manifest File: Uses-Featuretag

Within the AndroidManifest.xml file, various features are added apart from the sanctions. This information is applicable for the task of detecting malware.

```
<uses-feature
android:name="string"
android:required=["true" | "false"]
```

11.8 Results and Findings

Figure 11.2 shows the graphical user interface, which is a main form of our system. In this figure there are four major options, namely, scan file, package insert, rule list, and after rule insertion scan file option. Figure 11.3 shows 'insert all required android permission' option similar to the read permission, write permission, and find location permission. Figure 11.4 shows 'after adding the permission' update which deletes or cancels any unauthorized updates. Figure 11.5 shows 'after update' where a scan on the system is performed and if any match to the file is found then immediately message is displayed on the mobile screen that 'the application "xyz" looks like a malware. You should uninstall it'. Figure 11.6 shows 'insert the package name' event, which is used if a malware is not found by permission. In that case, the system is scanned with the help of package name. Figure 11.7 shows a display of the malware found according to the package name. If the malware does not match by permission then all files in the system are checked by package name, which includes

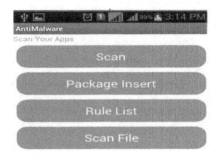

FIGURE 11.2
Setting permissions.

FIGURE 11.3
Applying permissions.

FIGURE 11.4
Malware detection – by permission.

FIGURE 11.5
Package insertion.

FIGURE 11.6
Insert the package name event

FIGURE 11.7
Malware found according to the package name

FIGURE 11.8
Malware detection – by signature.

both signed package name and unsigned package name. Figure 11.8 shows the check using unsigned package, which includes distrusted files or non-registered files.

Table 11.2 shows the results of the investigations, where a reduced time rule is observed by applying "by permission & by package" setup techniques. The average time for both the techniques observed is equivalent for most of the applications. Additionally, since our approach relies only on permissions and package setup, modifications on permissions do not harm our system.

TABLE 11.2

Results and Discussion

CLASSIFICTION	RULE SET/COMBINATION	Time by Rule (In second)
BY PERMISSION	SMS+CALL	Om 0.040
	BOOT+GPS	Om 0.089
	BOOT+SMS+GPS	Om 0.089
	BOOT+GPS/LBS+ CALL+INTERNET	Om 0.0147
	GET_TASK + READ_CONTACTS+ WAKE_LOCK GPS+SMS	Om 0.567
BY PACKAGE	DATA_SET	Om 0.566
BY SIGNATURE	YES "CN=Android Debug,O=Android,C=US"	0m 0.347

11.9 Conclusion

Here, we have described a new method for malware detection in Smart-phone's, which has been validated by installing different types of applications. It is based on permission analysis and distinguishes malware programs from benevolent programs. Furthermore, it is light-weighted and faster, which makes it additionally pertinent for smartphone platform requirements. We have examined different methods for improving malware tracking on smartphones. We have also presented a method of examining malware tracking "by signature" classification. Our future scope will be to broaden the analysis on benevolent programs in order to generate additional efficient malware detection methods based on autonomously edification and to detect malware which are unrelieved until now.

References

[1] A.D. Schmidt et al., "Detecting Symbian OS Malware through Static Function Call Analysis," *Proc. 4th Int'l Conf.Malicious and Unwanted Software* (Malware 09), IEEE, 2009, pp. 15-22.

[2] Schmidt, A.D., Peters, F., Lamour, F., Scheel, C., Camtepe, S.A., Albayrak, S.: Monitoring smartphones for anomaly detection. Mob. Netw. Appl. 14 (1) (2009), 92106

[3] D. Barrera et al., "A Methodology for Empirical Analysis of Permission-Based Security Models and Its Application to Android," Proc. 17th ACM Conf. Computer and Communications Security (CCS 10), ACM, 2010, pp. 73-84.

[4] I. Burguera, U.Z. Nadijm-Tehrani, S. Crowdroid. Behavior- Based Malware Detection System for Android. In: SPSM'11, ACM (October 2011).

[5] W. Enck et al., "A Study of Android Application Security," Proc. 20th Usenix Security Symp., Usenix, 2011; http://static.usenix.org/events/sec11/tech/full_papers/Enck.pdf.

[6] Yong Wang, Kevin Streff, and Sonell Raman, "Dakota State University Smartphone Security Challenges" 2012.

[7] La Polla, M., Martinelli, F., Sgandurra, D.: A survey on security for mobile devices. Communications Surveys Tutorials, IEEE PP(99) (2012) 1-26.

[8] William Enck, Machigar Ongtang, and Patrick McDaniel. "On Lightweight Mobile Phone Application Certification" The Pennsylvania State University, 2012.

[9] Abhijith Shastry, Murat Kantarcioglu, Yan Zhou, and Bhavani Thuraisingham. "Randomizing Smartphone Malware Profiles against Statistical Mining Techniques", 2013.

[10] Parmjit Kaur, Sumit Sharma, "Google Android A Mobile Platform: A Review." In *Recent Advances in Engineering and Computational Sciences (RAECS)*, 2014, pp. 1-5. IEEE, 2014.

[11] Tiwari Mohini, Srivastava Ashish Kumar and Gupta Nitesh" Review on Android and Smartphone Security" NRI Institute of Information Science and Technology, Bhopal,Madhya Pradesh, INDIA, *20143*, pp. 1-5. IEEE 2013.

[12] Parmjit Kaur, Sumit Sharma, "Literature Analysis on Malware Detection." In Recent Advances in Engineering and Computational Sciences (RAECS), *2014*, pp. 1-5. IEEE, 2012.

[13] Burguera, U.Z., Nadijm-Tehrani, S. Crowdroid: Behavior- Based Malware Detection System for Android. In: SPSM'11, ACM (October 2011).

[14] Yong Wang, Kevin Streff, and Sonell Raman, "Dakota State University Smartphone Security Challenges" 2012.

[15] La Polla, M., Martinelli, F., Sgandurra, D.: A survey on security for mobile devices. Communications Surveys Tutorials, IEEE PP(99) (2012) 1-26.

[16] William Enck, Machigar Ongtang, and Patrick McDaniel "On Lightweight Mobile Phone Application Certification". The Pennsylvania State University. 2011.

[17] Abhijith Shastry, Murat Kantarcioglu, Yan Zhou, and Bhavani Thuraisingham. "Randomizing Smartphone Malware Profiles against Statistical Mining Techniques" 2010.

[18] ParmjitKaur, Sumit Sharma, "Google Android A Mobile Platform: A Review." In *Recent Advances in Engineering and Computational Sciences (RAECS)*, 2014, pp.1-5. IEEE, 2014.

[19] TiwariMohini, Srivastava Ashish Kumar and Gupta Nitesh. "Review on Android and Smartphone Security" NRI Institute of Information Science and Technology, Bhopal, Madhya Pradesh, India, *20143*, pp. 1-5. IEEE 2013.

[20] Parmjit Kaur, Sumit Sharma, "Literature Analysis on Malware Detection." In Recent Advances in Engineering and ComputationalSciences (RAECS), 2014, pp. 1-5. IEEE, 2012.

[21] Burguera, U.Z., Nadijm-Tehrani, S. Crow droid: Behavior- Based Malware Detection System for Android. In: SPSM'11, ACM (October 2011).

[22] W. Enck et al., "A Study of Android Application Security," Proc. 20th Usenix Security Symp., Usenix, 2011; http://static.usenix.org/events/sec11/tech/full_papers/Enck.pdf.

[23] Yong Wang, Kevin Streff, and Sonell Raman, "Dakota State University Smartphone Security Challenges" 2012.

[24] La Polla, M., Martinelli, F., Sgandurra, D.: A survey on security for mobile devices. Communications Surveys Tutorials, IEEE PP (99) (2012) 1-26.

[25] William Enck, Machigar Ongtang, and Patrick McDaniel "On Lightweight Mobile Phone Application Certification" The Pennsylvania State University 2011.

[26] Abhijith Shastry, Murat Kantarcioglu, Yan Zhou, and Bhavani Thuraisingham, "Randomizing Smartphone Malware Profiles against Statistical Mining Techniques" 2010

Index